Syngas Production: Status and Potential for Implementation in Russian Industry

Vladimir Litvinenko · Bernd Meyer

Syngas Production: Status and Potential for Implementation in Russian Industry

Springer

Vladimir Litvinenko
Saint Petersburg Mining University
St. Petersburg
Russia

Bernd Meyer
Institute of Energy Process Engineering
TU Bergakademie Freiberg
Freiberg
Germany

ISBN 978-3-319-70962-8 ISBN 978-3-319-70963-5 (eBook)
https://doi.org/10.1007/978-3-319-70963-5

Library of Congress Control Number: 2017958020

Printed on acid-free paper

This Springer imprint is published by Springer Nature
The registered company is Springer International Publishing AG
The registered company address is: Gewerbestrasse 11, 6330 Cham, Switzerland

Contents

About the Authors

Prof. Vladimir Litvinenko Doctor of Engineering, Professor; Rector of the St. Petersburg Mining University since 1994.

Author of over 250 scientific papers including 5 monographs, 65 patents and inventions (Russian and foreign ones), Laureate of State Prize of the Russian Federation in Science and Technology for "Geological Atlas of Russia", Laureate of Russian Government Award in Science and Technology for "Development and implementation of environmentally friendly combined technologies of ore mining and complex ore processing that ensured putting into operation and development of the unique Yakovlevsky mine of rich iron ore" (2008), Laureate of St. Petersburg Government Prize for outstanding achievements in the field of higher and secondary vocational education (2008).

Recipient of state awards: the Order of Honor (1998), Order "For Merit to the Fatherland" 4th degree (2003) and 3rd degree (2010).

Chairperson from the Russian side of the Russian-German Resource Forum.

Member of the Presidential Commission on the development strategy of the fuel and energy complex and environmental safety; member of the Governmental Commission on the fuel and energy complex, mineral reserves replacement and increasing the energy efficiency of the economy.

Chairperson of the Governing Board of the International Competence Centre for Mining Engineering Education as a Category 2 Centre, under the auspices of UNESCO.

Honorary Doctor of the Technical University Mining Academy Freiberg since 2014.

Prof. Bernd Meyer Doctor of Engineering, Professor; Director of the Institute of Energy Process Engineering and Chemical Engineering (IEC) since 1994.

Professor Bernd Meyer is the author of over 400 scientific papers, including 10 monographs, 150 patents and inventions.

Elected president of Technical University Mining Academy Freiberg for two terms (2008–2015, maximum number of terms one can be elected as president for the university).

Laureate of Honorary Medal by the Association of German Engineers (Saxony) (2012). Recipient of a degree of honorary doctorate of National Mining University, Dnepropetrovsk, Ukraine (2012) and St. Petersburg Mining University, Russia (2016). Recipient of a degree of Honorary Professorship of Lomonosov Moscow State University, Russia (2015).

Honorary Scientist of the Russian Federation (2015, single foreigner to have received this honor from Ministry of Education and Science, Russia).

Initiator and Founding President of World Forum of Universities of Resources on Sustainability (2012–2015) and its Honorary Member (since 2015). Founder and Key Coordinator of the Competence Centre Deutsches Brennstoffinstitut: bergakademie. Founder and Key Coordinator of the Energy Raw Materials Network (ERN).

Co-founder of the Centre of Competence “Large Research Facilities for Efficient High Temperature Conversion Processes”.

Member of the Management Board and Scientific Speaker for the national platform “Regional Centre of Growth Innovative Brown Coal Integration in Middle Germany”.

Abbreviations

BGL	British Gas/Lurgi, fixed-bed liquid slag gasification technology
CAPEX	Capital expenditure
COD	Conversion of olefins to diesel
COORVED	CO_2-Reduktion durch innovatives Vergaserdesign, innovative gasification technology
DME	Dimethyl ether
ECUST	East China University of Science and Technology
FBDB	Festbett-Druckvergasung, fixed-bed dry ash gasification
FT	Fischer–Tropsch synthesis
GE	General Electric
GSP	Gaskombinat Schwarze Pumpe, entrained-flow gasification technology
GTI	Gas Technology Institute
HGT	Heavy gasoline treatment
HT-L	Hangtian-Lu, entrained-flow gasification technology
HTW	Hochtemperatur-Winkler-Vergasung, stationary fluidized-bed gasification
IGCC	Integrated gasification combined cycle
MHI	Mitsubishi Heavy Industries
MOGD	Mobil olefins to gasoline/distillate
MtG	Methanol-to-Gasoline
MTO	Methanol-to-Olefins
MTP	Methanol-to-Propylene
MtSynfuels	Methanol-to-Synfuels
OMB	Opposite multiple burner, entrained-flow gasification technology
OPEX	Operational expenditure
SNG	Synthetic natural gas

STF	Syngas-to-Fuel
TIGAS	Topsoe integrated gasoline synthesis
TPRI	Thermal Power Research Institute
TREMP	Topsoe recycle energy-efficient methanation process
TRIG	Transport integrated gasification

List of Figures

List of Tables

Chapter 1
Introduction

Coal is the fossil fuel with the most uniform distribution worldwide. But the coal quality differs significantly depending on the region. The suitability of coals for fuel or non-fuel use is basically governed by their characteristics. The key process for non-fuel coal utilization is syngas production by coal gasification. The selection of the most appropriate coal conversion pathway also depends on the local site conditions and the prospective markets for the products considered.

The present study focuses on the evaluation of economically feasible applications for high-ash coals from Russian coal districts. The most promising usc for such coals is the production of chemicals with high added value. This is mainly caused by the specific geographical position of the coal districts. They are located far away from large industrial centres. Thus, local demand for power production is relatively low and the specific transport costs should be minimized due to the long transport distances.

The study is structured as follows. Firstly, an overview of the basic gasification technologies is given, together with a discussion of the main requirements to select an appropriate coal for the given gasification technology. In addition, concepts for combined fuel and non-fuel coal utilization (Annex concepts) are presented. Such concepts enable significant cost savings in CAPEX and OPEX due to the synergy effects between a coal-fired power station and an affiliated coal chemistry plant. After that, an overview of the economically most important routes for syngas conversion will be given. The applicability of the technologies considered for the syngas production and utilization will be discussed taking into account technical and economic aspects. Finally, the most appropriate overall process chains were selected and subsequently used to calculate material and energy balances as well as economic parameters.

Finally, a general assessment is given of the potential application for non-fuel coal utilization under the specific economic conditions in Russia. The main focus is on high-ash coals, since the conversion of these coals into valuable products seems to be economically more feasible than directly selling the unrefined coal, due to the high specific transportation costs in the latter case.

V. Litvinenko and B. Meyer, *Syngas Production: Status and Potential for Implementation in Russian Industry*,
https://doi.org/10.1007/978-3-319-70963-5_1

Chapter 2
Investigated Coals

This study is based on the analysis data for ten Russian coal taken from four coal mining districts [1]. Two coals were selected as examples which represent the given broad range of coal characteristics: low-ash coal from the Taldinskaya mine and high-ash coal from the Komsomolets mine. The most important coal parameters with high relevance for the proper selection of the optimal gasification technology are given in Table 2.1.

Russia comes second place worldwide in terms of coal reserves (157 billion tonnes) [2]. The coal industry in Russia is a profitable branch, which was privatized and essentially modernized at the end of the 20th century. The Russian coal industry is operated entirely without public subsidies. The annual coal production is constantly on the increase, as shown in Fig. 2.1 (left) and approaching the production level achieved just before the collapse of the Soviet Union (1990: 395 Mt/a) [3]. The export percentage is relatively high, already exceeding 40%, with a rising tendency, see Fig. 2.1 (right).

The export orientation is even more pronounced for the biggest coal region in Russia, the Kuzbass coal district. As shown in Fig. 2.2, coal exports increased from 48% in 2008 to 60% in 2014.

The transportation distances from the Kuzbass district to the harbours in the Far East as well as to the potential markets in West Europe and in the European part of Russia are extraordinarily long; several thousand kilometres. Other coal districts in Russia are also located in remote areas and almost inaccessible regions (see Fig. 2.3).

V. Litvinenko and B. Meyer, *Syngas Production: Status and Potential for Implementation in Russian Industry*,
https://doi.org/10.1007/978-3-319-70963-5_2

Table 2.1 Characteristics of the selected coals [1]

Parameter	Unit	Taldinskaya	Komsomolets
Immediate analysis			
Fixer carbon	wt% (wf)	55.43	33.81
Volatile matter	wt% (wf)	30.97	30.66
Ash content	wt% (wf)	13.61	35.53
Water content	wt%	11.80	9.65
Elemental analysis			
C	wt% (wf)	70.60	53.47
H	wt% (wf)	3.97	3.82
O	wt% (wf)	9.31	4.96
N	wt% (wf)	2.15	1.59
S	wt% (wf)	0.22	0.55
Cl + P + As	wt% (wf)	0.15	0.10
Heating value (as received coal)			
Lower heating value H_i	MJ/kg	23.85	19.71
Higher heating value H_s	MJ/kg	24.90	20.69
Ash melting behaviour (reducing atmosphere)			
Softening temperature t_A	°C	1440	1310
Hemisphere temperature t_B	°C	1470	1360
Flow temperature t_C	°C	1500	1410
Further parameters			
Swelling Index	–	1	–
Hardgrove Index	–	58	57
Coal size distribution – Above 50 mm – Below 2 mm	mm % %	0–200 – –	0–200 11.1 20.4
Ash composition			
SiO_2 Al_2O_3 Fe_2O_3 CaO MgO TiO_2 MnO_2 P_2O_5 SO_3 Na_2O K_2O	wt% wt% wt% wt% wt% wt% wt% wt% wt% wt% wt%	60.90 23.40 5.00 2.90 2.00 0.90 0.03 0.40 1.30 0.8 2.40	42.66 16.75 10.13 10.58 3.86 0.89 0.03 0.20 9.06 2.18 2.63

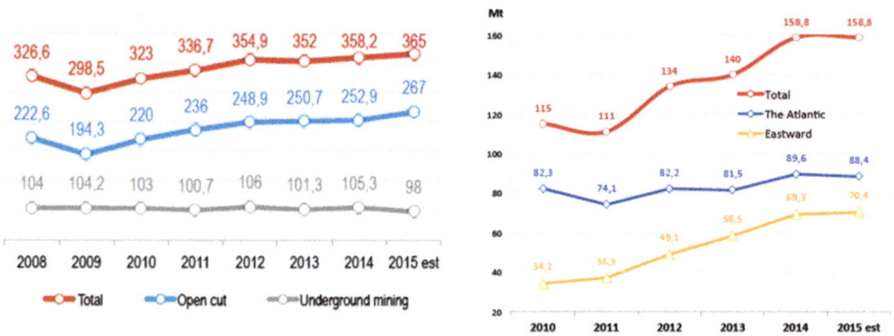

Fig. 2.1 Coal production (left) and coal export (right) in Russia [4]

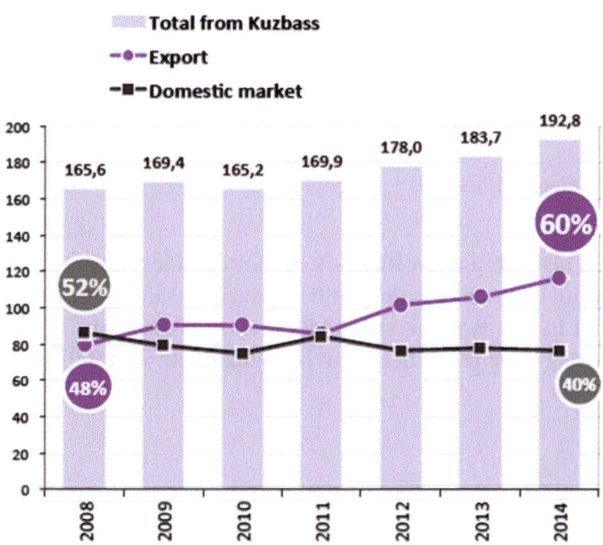

Fig. 2.2 Coal production in Kuzbass coal district [1]

A significant percentage of the coal deposits in Russia contain high-ash coals, characterized by a lower specific heating value. For such low-quality coals, the non-fuel use could be economically more feasible than selling the raw coal on the world market. The key technology in this respect is syngas production and utilization. The Russian coal industry's interest in the topic of coal gasification topic is steadily increasing. The implementation of advanced, economically feasible technologies for the coal-based production of valuable chemicals with high market

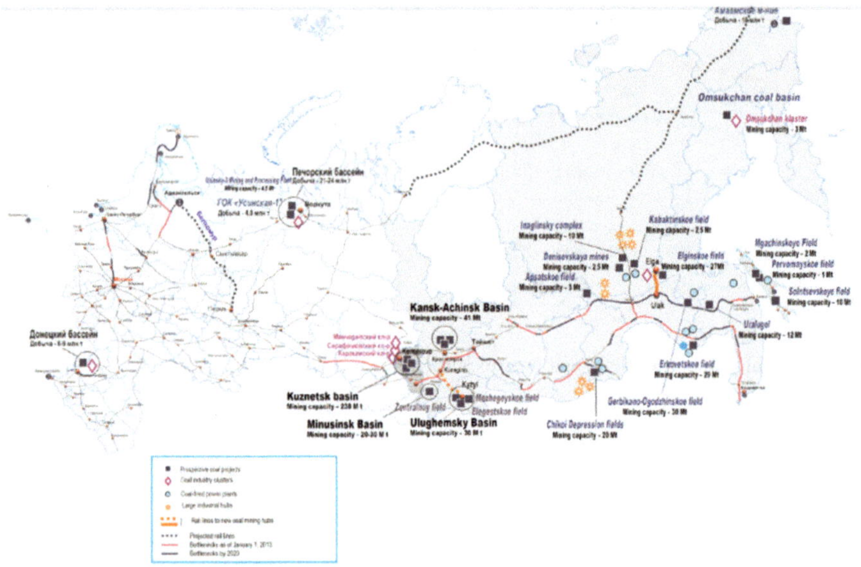

Fig. 2.3 Coal districts in Russia [4]

potential would also strengthen Russia's economic independency from the cyclical, market-driven fluctuations of commodities prices and install domestic value-added chains favoured by the Russian state. Moreover, existing jobs in the coal industry will be safe in the long term, and new high-qualified jobs will be created in coal chemistry clusters.

Chapter 3
Fundamentals of Gasification and Gasifiers

The gasification of Russian bituminous coal is the most suitable way to maximize the added value to this natural local resource. Coal gasification, the cleaning of the raw gas produced and its subsequent synthesis to create the final product are well-proven processes. For the most common final products (e.g. methanol, ammonia, synthetic fuels), each required process unit can be supplied by several vendors, each with a proven track record of successful commercial applications.

To select the coal gasifier, the most important task is to choose a gasifier that fits the respective coal. Not every gasifier is able to efficiently convert every coal quality. In the following section, the basics of coal gasification are briefly introduced. Thereafter, coal gasifiers are classified and described with a focus on their suitability to gasify highly volatile bituminous Taldinskaya and Komsomolets coals. It should be noted that Komsomolets coal has an ash content of 35.5 wt% (dry) and is categorised as high-ash coal.

3.1 Combustion and Gasification

In both cases, combustion and gasification, a coal particle is consumed via the following steps: heating, drying, devolatilization, consumption of the remaining coke. The remaining solid is ash [5].

The most important differences between combustion and gasification are the amount of oxygen that is added to the coal and the resulting composition of the generated gas. In a combustion process, enough oxygen is added (usually as air) to completely oxidize the hydrocarbons of the feedstock. As a result, heat is generated along with a gas composed mainly of nitrogen N_2 (in cases where air is used as an oxygen carrier) carbon dioxide CO_2 and water steam H_2O. In a coal-fired power station, the heat is used to generate power whilst the gas remains unused. In coal gasification, much less oxygen is added to the process (30–50% compared to combustion). Moreover, in most cases purified oxygen is used rather than air.

V. Litvinenko and B. Meyer, *Syngas Production: Status and Potential for Implementation in Russian Industry*,
https://doi.org/10.1007/978-3-319-70963-5_3

Because of the partial oxidation of the feedstock's hydrocarbons, a minor part of the feedstock's heating value is consumed to provide sensible heat. The generated gas is mainly composed of carbon monoxide CO and hydrogen H_2. In this raw gas, 70–85% of the coal's heating value are captured. It is possible to combust this gas in a gas turbine to generate power (IGCC power plant), but in most cases this gas is used a syngas to produce chemicals or fuels. Thus, by means of e.g. coal gasification, the feedstock (which is heterogeneously composed and difficult to transport and handle) is converted into a raw gas with a homogeneous composition that can easily be transported, further treated and exploited. The cleaned syngas can be utilized in manifold ways [6].

3.2 World Gasification Market

Currently, approx. 170 million tonnes of coal are gasified per year [7], which is half the amount of annual coal production in Russia. For the near future, the worldwide syngas capacity is forecasted to more than double from 147 GW_{th} (production capacity in 2015) to approx. 359 GW_{th} in 2020 [7]. Today, 65% of all the syngas produced is converted into chemicals (methanol and its non-fuel-derivatives,

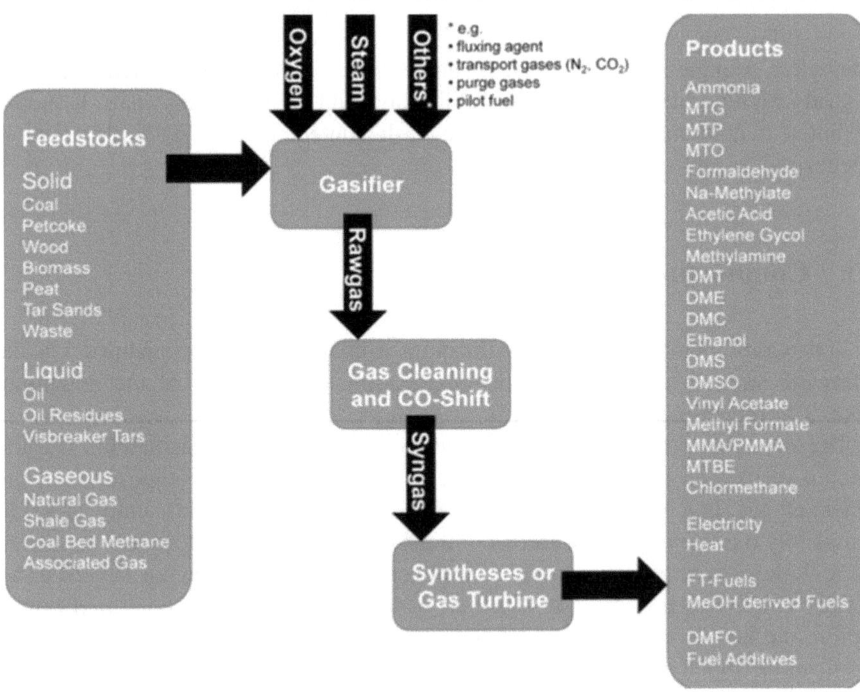

Fig. 3.1 Overview of gasification feedstock and syngas utilization

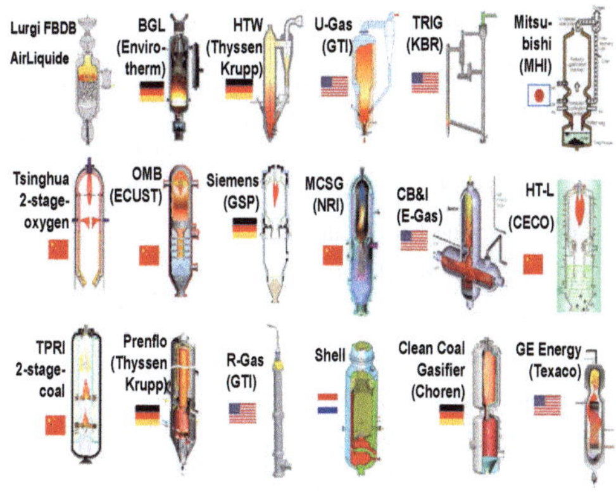

Fig. 3.2 Overview of commercial gasifiers

olefins, oxo-alcohols, ammonia, urea, hydrogen, etc.), 18% into liquid fuels (diesel, kerosene, naphtha, etc.), 10% into gaseous fuels (e.g. synthetic natural gas) and 7% into power (IGCC power plants, polygeneration). As shown in Fig. 3.1, there are many possible products and various applicable solid, liquid and gaseous feedstocks. However, 75.5% of the gasified feedstock is coal, and this percentage is actually increasing (83% by 2020).

There are many vendors for gasifiers. Figure 3.2 shows a few of them and Table 3.1 summarizes the market share of the most successful ones.

3.3 Coal Gasifiers

Since the beginning of industrial coal gasification in the early 20th century, many different types, ranks and qualities of coals have been converted into syngas. For this purpose, different types of gasifiers have evolved. These gasifiers can be classified according to various features, such as:

- bed type (fixed bed, fluidized bed, entrained flow)
- wall type (refractory lined, membrane wall, steam jacket)
- operational temperature (usually either above ash-fluid temperature or below the temperature of initial ash deformation, in a few cases in ash-agglomerating conditions)
- coal feeding (dry via lock hoppers, dry via solid pumps, slurry)
- raw gas cooling (heat exchangers, chemical quench, water quench)

Other distinguishing features include the operational pressure, oxygen purity, addition of fluxing agents and addition of catalysts. For this technical study, the bed type, operational temperature and raw gas composition are most interesting and will be further explained.

Table 3.1 Market share for most successful gasifiers [7]

Installed capacity in GW_{th} for different vendors	Operating	Under construction	Planned	Share in %	Share in %
	2014	2016	2019	2014	2019
GE (entrained flow, slurry)	30	43	46	25	19
Shell (entrained flow, dry)	30	32	39	25	16
Lurgi FBDB (fixed bed, dry)	19	20	21	16	9
OMB (entrained flow, slurry)	7	18	20	6	8
SEDIN (fixed bed, dry)	6	11	20	5	8
Siemens (entrained flow, dry)	2	14	17	2	7
E-Gas (entrained flow, slurry)	1	11	17	1	7
HT-L (entrained flow, dry)	2	9	13	2	5
MCSG (entrained flow, slurry)	5	5	6	4	3
Others	17	35	44	14	18

3.3.1 Counter Current Fixed-Bed Gasifiers

Commercial fixed-bed coal gasifiers are fed from the top and the gasification agent flows in a counter-current direction from the bottom to the top. The highest oxygen concentration is found at the very bottom. Here, an oxidation zone forms in which all the oxygen is consumed. Above this, endothermic conversion of devolatized coke takes place in the gasification zone. The gas produced flows upwards and provides heat for the devolatilization and drying of coal in the two zones above. This internal integration of combustion heat results in lowest oxygen consumption and lowest raw-gas outlet temperature (350–800 °C) compared to other gasifier types. In fact, this makes fixed-bed gasifiers most suitable for high-ash coals if the particle size is feasible. Due to the low gas outlet temperatures, noticeable amounts of methane and higher hydrocarbons (tars, oils) can be found in the raw gas (compare Table 3.2). Thus, the syngas yield is lower compared to fluidized-bed and entrained-flow gasifiers, even though the cold-gas efficiency is superior. The gas composition is especially suitable for the production of synthetic natural gas (SNG) but the production of methanol and its daughter products is inefficient. However, the solid bottom product is virtually free of carbon. It is discharged as dry ash by means of a rotating grate and rabble rakes (e.g. Lurgi fixed-bed dry bottom (FBDB) gasifier, SEDIN gasifier) [8], or as liquid slag through a tap hole (British Gas/Lurgi (BGL) gasifier) [9]. All fixed-bed gasifiers are dry fed with the

Table 3.2 Typical gas composition (vol.%, dry) from fixed-bed gasifiers [6]

Gasifier	CO	H_2	CO_2	CH_4	N_2, Ar, C_xH_y, etc.
Lurgi FBDB®	20	40	27	10	3
BGL	55	28	5	7	5

comparably largest coal particle size, from 3 to 60 mm. Fines must be removed during the feedstock preparation by means of sieving. Fines would result in an instable gas flow through the particle bed. The consequences would be an unequal distribution of reactant gas over the bed diameter and prohibitive fluctuations in pressure. The former results in an incomplete carbon conversion and hot spots in the packed bed. The latter results in random gas eruptions at the top of the bed and an accompanying increase in particle carry-over. In general, fixed-bed gasifiers have a comparably low oxygen consumption and a high steam consumption. Only little effort is required for coal grinding or crushing. The feedstock addressed in this technical report can be gasified in fixed-bed gasifiers if their caking index is low. However, the resulting gas composition causes poor economics if methanol is produced (see also Sect. 5.2.2). Thus, fixed-bed gasification cannot be recommended.

3.3.2 Fluidized-Bed Gasifiers

Circulating fluidized-bed gasification is offered by Foster Wheeler, Outotec or Envirotherm, among others. These systems operate only at near atmospheric pressures and have hardly any successful commercial demonstration of coal gasification. Stationary fluidized-bed gasification is offered by ThyssenKrupp (HTW gasifier). Fast fluidized-bed coal gasification is offered by KBR. Here, operational principles from fluid catalytic cracking (FCC) technology are adapted. Jetting fluidized-bed coal gasification is offered by SES. Regardless of the type of fluidization, all fluidized-bed gasifiers are dry fed with particle sizes between 1 and 6 mm. They have a dry-ash discharge and a moderate consumption of steam and oxygen. Typical gas outlet temperatures are between 800–1000 °C. Because of the moderate temperatures, fluidized-bed gasifiers have a thermodynamic advantage compared to entrained-flow gasifiers. This becomes especially evident for high-ash coals [6] as neither the heat of ash fusion, nor the sensible heat for heating the ash to higher temperatures is required. Thus, fluidized beds are very promising for gasifying coals with a high ash content or a high ash melting temperature. Nonetheless, fluidized-bed gasifiers strongly lag behind fixed-bed gasifiers and entrained-flow gasifiers in terms of installed capacity. In fact, there are only very few commercial demonstrations of coal gasification in fluidized beds. The major drawback is the incomplete carbon conversion of 90–95% only. This is a result of the residence time distribution and solids back-mixing in this reactor type. As a consequence, both the bottom ash (up to 5–25 wt% C content) and fly ash (up to 50–60 wt% C content) contain considerable amounts of unconverted carbon. It is necessary to post-combust these carbon-rich residues in a subsequent combustor. Usually, this is done in a nearby power station. Moreover, fluidized-bed gasifiers need a highly reactive coal (lignite, sub-bituminous) to achieve 90–95% carbon conversion [10]. Bituminous coals have a carbon conversion of 60–80% only. The moderate gas outlet temperatures from the gasifier result in a methane content of 3–6%, which

Table 3.3 Typical gas composition (vol.%, dry) from fluidized-bed gasifiers [6]

Gasifier	CO	H_2	CO_2	CH_4	N_2, Ar, C_xH_y, …
HTW	37	34	19	4	6

would reduce the product yield of methanol from the syngas in a subsequent synthesis (compare Table 3.3). The investigated Taldinskaya and Komsomolets coals cannot be recommended for gasification in a fluidized-bed gasifier. The carbon conversion rate would be very low; post-combustion of the ash would be required. Both result in poor economics.

3.3.3 *Entrained-Flow Gasifiers*

Apart from fixed-bed gasifiers with a dry ash discharge, only entrained-flow gasifiers make up a relevant market share. Especially in the last couple of years, entrained-flow gasifiers have mainly been applied. The reasons are their complete carbon conversion, a raw gas free of higher hydrocarbons and a very low methane content (compare Table 3.4). The most important feature of entrained-flow gasifiers is the very high single unit capacity. Most vendors offer single units with a coal consumption capacity of 1500–2000 t/d. The Chinese OMB gasifier is currently operated on a 3000-tpd scale (Inner Mongolia, Rongxin, 65 bars, slurry gasifier) [11]. The US KBR gasifier is available as a 5000-tpd unit, which means approx. 1000 MW of thermal input [12]. On the other hand, the required effort for coal preparation is comparably high. Coal needs to be ground to particle sizes below 0.25 mm and fed dry or as a slurry. Dry feeding requires a certain amount of transport gas (usually N_2) which affects the syngas quality. Slurry feeding requires a certain amount of water, which reduces the cold gas efficiency of the process. Compared to fixed-bed and fluidized-bed gasifiers, entrained-flow gasifiers have a lower residence time (2–5 s). In order to achieve complete particle burnout, temperatures above 1200 °C are required. The oxygen is fed through at least one burner. Thus, a hot flame of more than 2000 °C is formed. Due to the hot flame and the high mean temperature, the particle ash starts to liquefy. A certain ash liquid temperature must be exceeded all over the reaction chamber. In this respect, accurate knowledge of the ash liquid temperature is essential for operation. Depending on the coal, this temperature is between 1200 and 1600 °C. Usually, the

Table 3.4 Typical gas composition (vol.%, dry) from entrained-flow gasifiers [6]

Gasifier	CO	H_2	CO_2	CH_4	N_2, Ar, C_xH_y, …
Siemens (dry feed)	60	29	5	0	6
GE (slurry feed)	41	34	23	0	2

gasifier is operated 100–150 K above this value for safety reasons [6]. Thus, entrained-flow gasifiers have a raw gas outlet temperature between 1350 and 1700 ° C. In this way, the liquefied ash (slag) can flow down the reactor walls smoothly and finally drop into a water bath, where it is solidified. The need to melt the ash and to reach the required operational temperatures is especially disadvantageous for high-ash coals. Even though it is technically feasible to gasify high-ash coals, the process has a poor thermodynamic performance and questionable economics [13]. In particular, slurry-fed gasifiers should not be operated on high-ash coals as the slurry water and the ash need to be heated up to the high reactor temperatures. However, dry-fed entrained-flow gasifiers can provide complete carbon conversion and a very good syngas quality if Taldinskaya or Komsomolets coals are gasified. The high-ash Komsomolets coal would cause a high oxygen consumption and a gas of slightly lower quality (increased CO_2 content).

All entrained-flow gasifiers have a higher oxygen consumption and lower steam consumption compared to fixed-bed and fluidized-bed gasifiers. Steam is required for temperature moderation if the coal has a high heating value, which would result in too high temperatures if converted with oxygen and would consequently damage the reactor materials. Low-rank coals might be gasified even without moderator steam. Also, slurry-fed gasifiers do not apply additional steam for moderation. Another distinguishing feature is the applied wall type. There are refractory lined walls and membrane walls. Refractory lined walls are cheaper to construct but they have a limited lifetime (between six months and one year). Thermal, mechanical and chemical stresses require the periodic replacement of damaged bricks. On the other hand, the glowing bricks serve as a permanent source of ignition if, for some reason, the coal-oxygen flame is interrupted. Membrane walls are more expensive but can be operated for many years (usually, vendors guarantee 10 years, but 15–20 years have been achieved). The membrane wall in a Siemens gasifier (compare Fig. 3.3) features a spirally wound coil covered with a SiC castable. Hot, liquid slag that reaches the cooled wall solidifies on the castable and forms a protective solid layer [5]. Depending on the heat transport properties of that layer, it grows to a certain thickness. Liquid slag flows down on that solid slag layer. In this way the reaction chamber is protected from slag intrusion, spalling and thermal stresses, whilst the layer has a self-healing tendency if holes occur in the layer. Membrane walls allow the start-up and shutdown procedures to be fast, which increases their availability compared to refractory lined gasifiers. On the other hand, there is no glowing surface serving as an ignition source. Thus, the Siemens gasifier, for example, features a pilot burner (compare Fig. 3.3) that backs up the coal-oxygen flame.

Entrained-flow gasifiers can be equipped with various raw-gas cooling units. The cheapest and most robust option is a full water quench as offered e.g. by GE, Siemens, Shell or ThyssenKrupp. The hot raw gas is quenched in a spray quench or dip tube unit to temperatures between 180 and 230 °C. In the same quench chamber, slag drops into the bottom water bath, solidifies and is extracted. The cooled and steam saturated gas is directed to the downstream units. The thermal

Fig. 3.3 Siemens gasifier

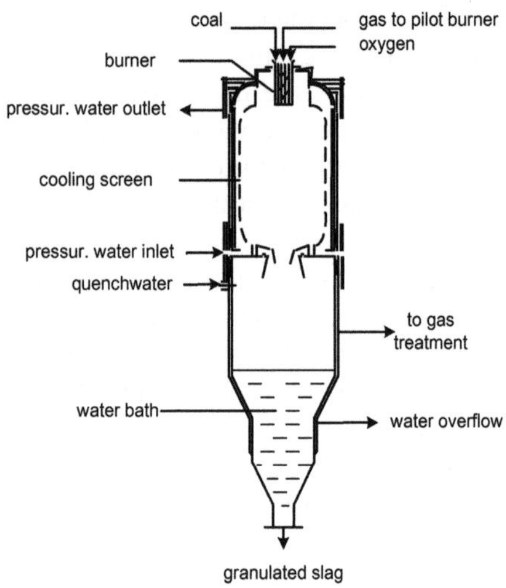

efficiency of a quench gasifier is lower compared to a system with heat recovery. Heat is recovered from the hot raw gas via radiative and/or convective heat exchangers. In this configuration, the production of high-pressure and/or medium-pressure steam is possible. Heat exchanger steam generation improves the efficiency but increases the CAPEX and reduces the plant availability, as the heat exchanger surfaces are prone to fouling and degradation.

3.3.4 Optimized Gasifier with a Staged Bed (COORVED Gasifier)

The superior heat integration and low investment costs of a fixed-bed gasifier can be combined with the moderate steam and oxygen consumption of a fluidized-bed gasifier and the gas quality of an entrained-flow gasifier. This is possible in a staged gasifier, comprising a fixed bed at the bottom and a jetting fluidized bed above [14]. This concept was successfully demonstrated at the Technical University in Freiberg, Germany. The COORVED gasifier is characterized by increasing gas velocities from bottom to top. Figure 3.4 shows how the gasifier is composed of four subsequent zones. Zone 1 is a fixed bed at the bottom of the reaction chamber. In this zone, char agglomerates (that are formed mainly in the jetting bed) are contacted with a secondary gasification agent. The secondary gasification agent is a mixture of oxygen and steam and is introduced via a rotating grate, comparable to a Lurgi fixed-bed gasifier. In addition, this grate is used for discharging the bottom ash into the ash sluice. Ash discharge can easily be controlled by the rotational speed of the grate. Above the fixed bed, a tapered segregation zone can be found.

Fig. 3.4 COORVED gasifier

Here, particles are separated according to their weight. Larger particles (agglomerates, large feedstock particles) move downwards and fall onto the bottom fixed bed. Smaller particles are directed upwards into the main reaction zones (jetting fluidized bed and fast fluidized bed). Only a small size fraction will reside permanently in a fluidized bed in this region. However, if necessary, the separation effect can be adjusted by changing the reactor diameter in this region. A jetting fluidized bed can be found above the tapered region. This is the main reaction zone. Oxygen is added as the primary gasification agent through nozzles. Due to the high outlet velocity at the nozzle mouth, a jet emerges and penetrates the fluidized bed. The fresh feedstock enters here and comes into contact with the primary gasification agent, which leads to dominantly exothermic reactions. Thus, the jetting bed has the highest temperature in the reactor. In fact, a flame-like hot zone of more than 2000 °C is formed. If particles reach a certain carbon conversion, their ash surface will begin to partially soften or melt due to the high temperatures. In this way, particle surfaces become sticky, which allows agglomerates to form. The agglomerates are too heavy to be fluidized. Along with large feedstock particles, they fall through the bubbling bed and form the fixed bed at the bottom. In the main reaction zone, particles are dried and devolatized. Because the temperatures are comparably high, devolatilization gases such as CH_4 and higher hydrocarbons are destroyed. Thus, the produced gas quality is comparable to an entrained-flow gasifier. The upmost region above the jetting bed has the highest gas velocity because gas is released as the feedstock conversion progresses. The resulting fast-fluidized bed is operated between the conditions of a circulating fluidized bed and entrained-flow conditions. As there is no more oxygen in this upmost zone, only endothermic reactions occur.

The gas temperature is decreased to 1000–1200 °C. If required, the gas outlet temperature can be further reduced by means of partial water quenching in the head of the reactor. In this way, gas outlet temperatures of 900–950 °C can be achieved [14].

Due to the moderate gas outlet temperature, the cold gas efficiency and syngas yield is superior to other bed types. In addition, the gas is very poor in methane, as the gasifier features a hot flame (>2000 °C) in the jetting fluidized bed. In addition, the COORVED gasifier can be operated over a broad particle size range for the feedstock (0–6 mm). Thus, this gasifier shows a superior performance for all types of coal. For ash-rich coal (e.g. Komsomolets coal), the performance is much better compared to entrained-flow gasifiers, as there is no need to melt the ash completely.

The COORVED gasifier is designed to require little investment and cost little to operate. In addition, its performance is superior to entrained-flow and fixed-bed gasification in terms of the syngas yield and quality. Thus, the COORVED gasifier incurs low production costs for chemical products such as methanol (see also Chap. 7).

3.4 Selection of Gasification Technologies

Considering technological and economic aspects, the following two gasifiers are chosen for this report:

- Siemens entrained-flow gasifiers
- COORVED staged fluidized-bed gasifier

Chapter 5 shows that the most feasible products of syngas utilization that fit Russian coals are methanol (a basic feedstock for the chemical industry) and gasoline. These targeted products can be efficiently synthesized from a raw gas with minimal methane content. Thus, fixed-bed gasifiers cannot be recommended, as they produce up to 12% methane. The comparably high ash-softening temperatures would allow gasification to take place in a conventional fluidized bed. On the other hand, it has been shown that bituminous coals have a low degree of carbon conversion (<85%) in fluidized beds and thus unacceptable economics. The most promising commercial gasifiers for the coals considered and products desired are entrained-flow gasifiers. Among them, dry-fed systems are more favourable for high-ash coals such as Komsomolets coal. It is known that the Siemens gasifier has a much lower CAPEX compared to the Shell and Prenflo gasifiers. In this respect, the Siemens gasifier is a reasonable choice for further evaluation. Nevertheless, entrained-flow gasifiers have the highest oxygen consumption and only intermediate cold gas efficiencies. If a coal with a high ash content is applied, the key performance indicators suffer strongly. Hence, entrained-flow gasifiers have generally not been applied for the utilization of such coals. Thus, a second gasifier is evaluated in this report. The new COORVED gasifier was developed to be especially suitable for high-ash and low-grade coals. It is characterized by a complete carbon conversion accompanied by low oxygen consumption and a very good syngas quality. Even for low-ash coal, the performance is competitive compared to entrained-flow gasifiers. For high-ash coals (e.g. Komsomolets), the performance is superior.

Chapter 4
Annex Concepts

Coal gasification with subsequent syngas conversion is a mature technology. The main challenge today is minimizing the plant CAPEX. In contrast to oil-based process chains, coal utilization requires expensive facilities for the coal preparation, syngas cleaning and wastewater processing. Investigations at the Institute of Energy Process Engineering and Chemical Engineering show that the CAPEX can be significantly reduced thanks to a special coupling between the fuel- and non-fuel-based coal utilization processes.

4.1 Concept Description

A commercial-scale coal gasifier is installed next to a coal-fired power station, the former having a lower thermal capacity than the latter. This process chain is referred to as the "Annex". The chemical plant include the coal preparation facility, air separation unit, gas cleaning unit and synthesis block. The power station and Annex plant are connected by several mass streams. The excess steam (remaining after all internal steam consumers have been supplied), waste gases, wastewater and carbon-containing solid residues are forwarded from the Annex plant into the power station. In the opposite direction, the Annex plant serves as a permanent power consumer for the coal-fired power station. If required, steam at a certain pressure level could be provided. The general scheme for the Annex concept is shown in Fig. 4.1.

The Annex plant can also include a water electrolyser. The hydrogen produced can be integrated into the synthesis process. As a consequence, the CO-shift reactor can be diminished in size and the product yield can be increased. The benefits for the power station are a higher power demand thanks to the operation of the electrolyser and increased material streams forwarded from the Annex plant to the power station.

V. Litvinenko and B. Meyer, *Syngas Production: Status and Potential for Implementation in Russian Industry*,
https://doi.org/10.1007/978-3-319-70963-5_4

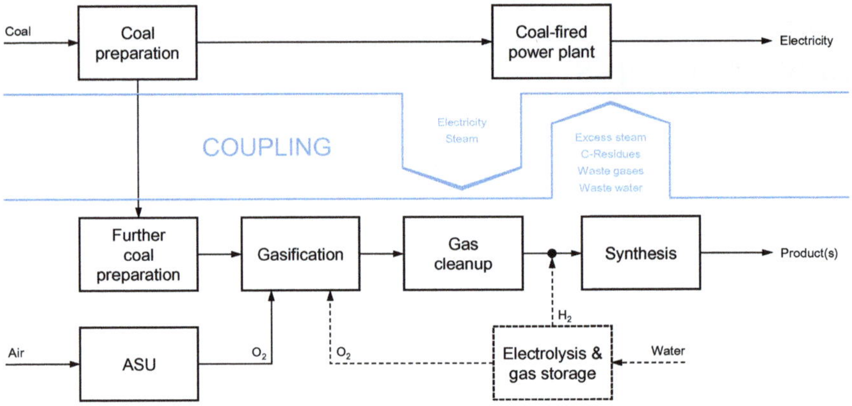

Fig. 4.1 General scheme for the Annex concept

4.2 Technical and Economic Benefits

There are several technical and economic benefits resulting from a strong interconnection between the power station and the Annex plant. The synergies of both process chains realized at one production site allow CAPEX costs to be saved, because some of the components in the Annex plant can be omitted. The post-treatment of the off-gases from the Claus unit is no longer necessary; these gases are disposed of in the power station boiler together with heating gases from the synthesis unit (light ends, flare gases). Pollutants from the combustion process, such as SO_2, are removed by the standard flue gas desulphurisation of the power station. Carbon-containing solid wastes (e.g. from the fluidized-bed gasification) could be also disposed of in the power station boiler. Thus, there is no need for a separate reactor for the post-combustion of the carbon-rich ashes. The ash produced can be dumped in accordance with environmental protection regulations. Wastewater from the gas cleaning processes (grey water, black water) often features a high solids load. It can be concentrated and also disposed of in the power station boiler. Remaining wastewater, especially from the synthesis process, could be forwarded to the power station wastewater treatment unit. The feed water for the steam generators in the Annex plant is supplied by the power station. In addition, the coal preparation and storage facilities of the power station could also be used for the Annex plant.

These aspects offer significant cost savings regarding both the CAPEX and OPEX of the Annex plant. Estimations made by the authors of this study show that CAPEX costs can be reduced by 15–21% depending on the relative dimensions of the power station and the gasifier, as well as the selected gasification technology and subsequent synthesis processes.

In practice, the actual boundary conditions at the existing power stations should be taken into consideration. If necessary, some compromises could be made with respect to the extent of the interconnections between the power station and the Annex plant. In the case of a newly constructed power station, the possible interfaces can be adapted more accurately.

Some advantages also result for the power station, especially in terms of the operational mode. This applies preferentially for the European electricity market and will be illustrated here using lignite-fired power stations in Germany as an example. Originally, these power stations were designed for electricity generation in the basic load mode. Recently, they have needed to be operated in a more flexible mode due to the increasing fluctuations in the power grid caused by the growth of renewable power generation. In the case of excessive electricity generation by wind power stations and solar plants, conventional coal-fired power stations must reduce their load power. The critical criterion is the minimal load power. The coal-fired power station cannot operate below this level due to certain technical limitations (boiler operation conditions, steam cycle requirements). If necessary, coal-fired power stations should be shut down, resulting in a lower annual availability at the expense of economics. Frequent start-up and shutdown procedures reduce the lifetime of basic components of the power stations. The installation of an Annex plant provides a permanent electricity consumer and help to reduce the minimum load power of the coal-fired power station in terms of the electrical grid. Therefore, the power station requires fewer shutdown phases and the annual availability can be increased. In addition, the mass flows from the Annex plant result in lower OPEX costs for the steam generators at the power station, reduced coal demand and consequently decreased CO_2 emissions.

In contrast to the power station, which can be operated in a flexible mode, the Annex plant should run under stable process conditions. The effect of reducing the minimum load power of the coal-fired power station can be additionally strengthened using electrolysers in the Annex plant, which can be quickly put into operation if necessary. The benefits in terms of the Annex plant were already mentioned above, see Sect. 4.1.

As an example, a technical study performed at the Institute of Energy Process Engineering and Chemical Engineering and funded by German coal companies shows that the following benefits were estimated for an integrated plant complex, consisting of an existing lignite-fired power station with an electrical power of 600 MW(el), an Annex plant for a Coal-to-Liquids process with a thermal power of 200 MW(th) and an electrolysis unit with 50 MW(el):

- Increase in the product yield by more than 26%
- Decrease in the minimum load power of the power station by more than 22%
- Reduction in the product costs by approximately 16%
- Reduction in CO_2 emissions for the Annex plant and the power station
- Mass flows can be integrated into the power station without serious impacts.

Chapter 5
Syngas Utilization Technologies

5.1 Overview of Syngas Utilization Routes

Syngas can be used for the production of numerous chemicals, as shown schematically in Fig. 5.1.

One of the key products is methanol, the syngas-based formation of methanol proceeds according to the following equation:

$$CO + 2H_2 \rightarrow CH_3OH \tag{5.1}$$

The conversion of syngas into methanol is described in detail in Sect. 5.2.2. Methanol is one of the most important raw materials used for the production of several chemicals, e.g. formaldehyde, acetic acid, ethylene and propylene. The conversion of methanol into ethylene and propylene, the monomers for the polymer production, is described in Sect. 5.2.4. A further conversion route is the production of synthetic liquid fuels. The methanol-based process for fuel production proceeds according to the equation:

$$CO + 2H_2 \rightarrow CH_3OH \rightarrow [-CH_2-] + H_2O \tag{5.2}$$

Here [–CH_2–] symbolically stand for the produced mixture of hydrocarbons, consisting of saturated, unsaturated and aromatic compounds. To convert methanol into fuels, one-step processes (see Sect. 5.2.3) and multiple-step processes (see Sect. 5.2.5) have been developed.

An alternative route for the syngas utilization is Fischer–Tropsch synthesis (FT synthesis), described in Sect. 5.3. FT synthesis is a complex chemical process; the reaction mechanism is not yet known. The simplified overall reaction equation can be given as follows:

V. Litvinenko and B. Meyer, *Syngas Production: Status and Potential for Implementation in Russian Industry*,
https://doi.org/10.1007/978-3-319-70963-5_5

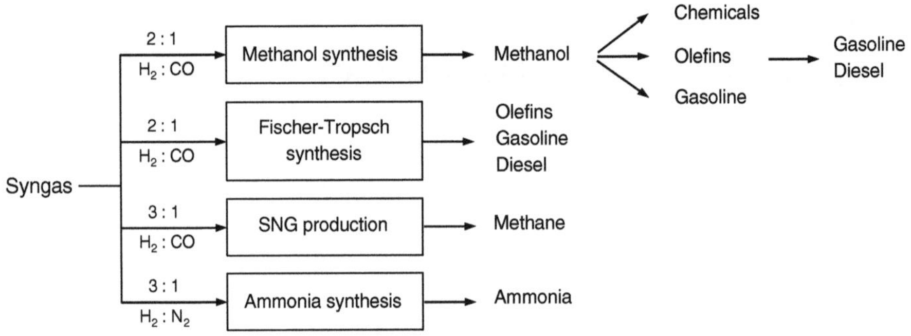

Fig. 5.1 Syngas utilization routes

$$CO + 2H_2 \rightarrow [-CH_2-] + H_2O \tag{5.3}$$

The produced mixture of hydrocarbons reveals a very broad product spectrum with compounds containing up to 50 C atoms. The basic difference between the methanol-based route for fuel production and FT synthesis consists in the product selectivity. An economic comparison of these routes shows that CAPEX and OPEX for both routes are comparable [15], though only concerning the formation of the raw product. In contrast to the methanol-based process, FT synthesis requires complex and expensive post-treatment of the raw product in a refinery.

The coal-based syngas can be converted into methane:

$$CO + 3H_2 \rightarrow CH_4 + H_2O \tag{5.4}$$

The methane produced is referred to as Synthetic Natural Gas (SNG). The process is described more extensively in Sect. 5.4. It should be noted that the production costs for coal-based SNG are not competitive with natural gas. This conversion route could be economically feasible for regions with large coal deposits and poor infrastructure for a natural gas supply.

Hydrogen produced by coal gasification can be used to produce ammonia:

$$N_2 + 3H_2 \rightarrow 2NH_3 \tag{5.5}$$

Ammonia is the most important feedstock for the production of fertilizers and chemicals (see Sect. 5.5). Coal-based ammonia production is especially common in China.

5.2 Methanol and Subsequent Products

5.2.1 *Overview*

Methanol is one of the most important feedstocks for the chemical industry. An overview of the application areas of methanol is given in Fig. 5.2.

5.2.2 *Methanol Production*

5.2.2.1 Market Analysis

Worldwide methanol production is 76 million t/a (in 2016), more than 95% of which is produced from syngas [16]. Syngas-based methanol production is a standard, well-developed process. The most important licensers for methanol synthesis are Lurgi (27% market share), Johnson Matthey/Davy (25%) and Haldor Topsøe (16%) [17]. Further vendors are the Mitsubishi Gas Chemical Company, Johnson Matthey/Uhde, Johnson Matthey/Toyo and others. Once, syngas was mainly produced from natural gas. In recent years, the share of the coal-based route for methanol production has constantly increased to approximately 40% [18]. Methanol production from coal is especially common in China. Some methanol synthesis plants using coal-based syngas are listed in Table 5.1.

Methanol prices in Europe and Asia in the last 10 years are given in Fig. 5.3. Methanol is one of the most important basic chemicals used to produce various compounds. The distribution of the methanol consumption is shown in Fig. 5.4.

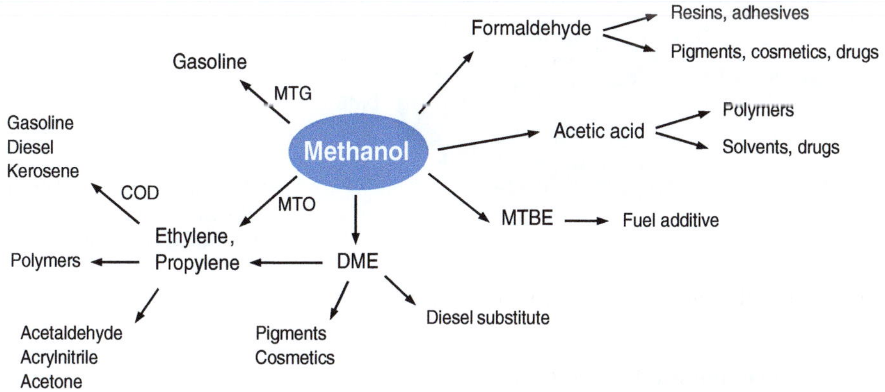

Fig. 5.2 Application areas of methanol

Table 5.1 Selected coal-based methanol plants

Plant	Country	Year	Technology	Capacity (in kt/a)	Lit.
Zhongyuan Dahua	China	2007	Haldor Topsøe	470	[19]
Tianjin Soda	China	2009	Haldor Topsøe	470	[19]
Xianyang, Shaanxi	China	2009	Haldor Topsøe	660	[19]
Shenhua, Baotou	China	2010	Johnson Matthey/Davy	1800	[20]
Zhong Tian Energy	China	2011	Lurgi	3600	[21]
China Blue Chemical	China	2011	Johnson Matthey/Davy	2500	[22]
Datang Duolin Coal	China	2012	Lurgi	1800	[23]
Sinopec/Wanbei	China	2013	Lurgi	1800	[21]
Ningbo Heyuan	China	2013	Johnson Matthey/Davy	1800	[23]
Shenhua Ningxia Coal	China	2014	Lurgi	1800	[23]

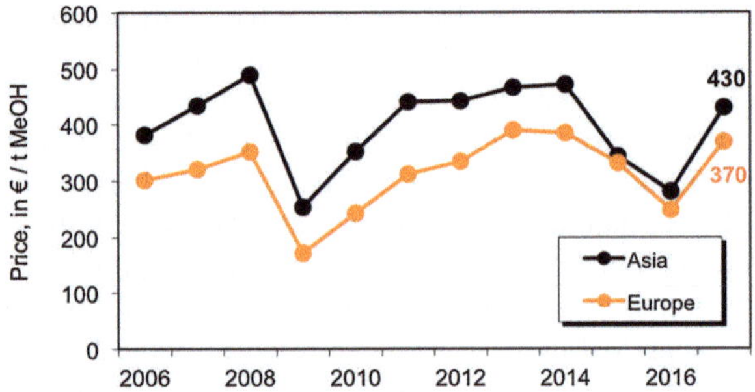

Fig. 5.3 Methanol prices in Europe and Asia [24]

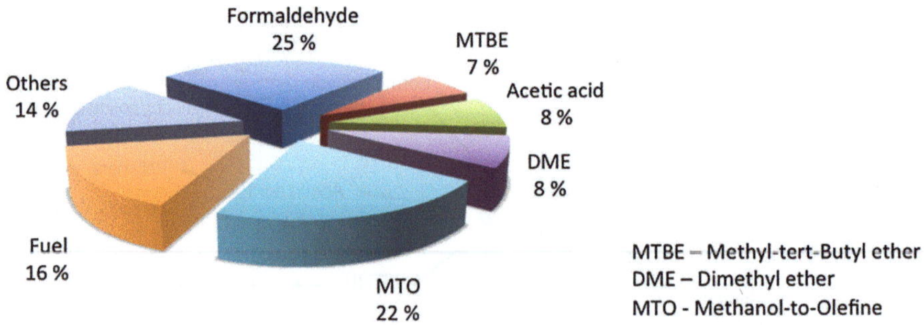

Fig. 5.4 Distribution of methanol consumption [25]

5.2.2.2 Technology Description

Methanol has been produced from syngas as in Eq. 5.1 (see Sect. 5.1) on a commercial scale for almost 100 years. Former high-pressure methanol synthesis technologies (250–350 bar, 320–400 °C, ZnO catalyst) have been replaced, thanks to the development of more efficient catalysts, by low-pressure synthesis (50–100 bar, 230–280 °C, $Cu/ZnO/Al_2O_3$ catalyst) [17]. Due to the chemical equilibrium, syngas conversion to methanol is incomplete after a single reactor pass. Typically, the methanol yield after a single reactor pass is approximately 50% [17]. The equilibrium is shifted towards methanol as a product at higher pressures and lower temperatures. The unreacted syngas is recycled back to the methanol reactor. The typical ratio between the recycle gas flow and feed gas flow varies from 3:1 to 7:1 [26]. The higher recycle ratio is attributed to a higher product yield, but also to a higher energy demand for the compressors. In order to avoid an accumulation of inert components (nitrogen, noble gases, methane, etc.) in the recycle gas, part of the gas stream (usually approximately 5–10%) is discharged as purge gas. From the thermodynamic point of view, methane is the more favourable product under the reaction conditions of the methanol synthesis. However, methanol production is favoured thanks to the chosen catalyst. Commercial copper-based catalysts (e.g. from Johnson Matthey) enable a very high selectivity of the methanol formation, which usually exceeds 99% [17].

5.2.2.3 Fixed-Bed Processes

Gas-phase synthesis technologies using fixed-bed methanol converters are available for the methanol synthesis. The first low-pressure process for methanol synthesis was developed by Imperial Chemical Industries (ICI, today Johnson Matthey), using adiabatic methanol converters. Such converters possess several catalyst layers stored on sieve plates. The cold syngas is inserted between the catalyst layers using special gas distribution systems. The thermal efficiency of the adiabatic methanol synthesis is lower than for quasi-isothermal synthesis processes. On the other hand, adiabatic processes are characterized by a less expensive construction of the methanol converter (without heat exchangers inside the reactor) and therefore by lower CAPEX.

Most installed methanol plants use the Lurgi process with quasi-isothermal tubular reactors. Methanol is produced at 240–260 °C, 60–80 bar, with commercial $Cu/ZnO/Al_2O_3$ catalysts (e.g. from Clariant) [17, 27]. Slim tubes filled with a catalyst are immersed in boiling water to remove the exothermic reaction heat. The temperature inside the converter is regulated by controlling the pressure in the steam drums. The quasi-isothermal process enables a higher methanol yield after a single reactor pass due to the more uniform temperature distribution inside the methanol converter. In addition, steam generation allows more efficient heat utilization. The gas mixture is cooled down at the reactor outlet under the formation of a methanol-water-rich condensate. This raw methanol product is treated in a distillation column to produce pure methanol.

Methanol plants with a high annual output (>1 million t/a) generally use two-stage synthesis technologies, e.g. the "MegaMethanol®" process by Lurgi [28]. These technologies are characterized by lower specific OPEX and higher thermal efficiency. The "MegaMethanol®" process is based on the combination of a

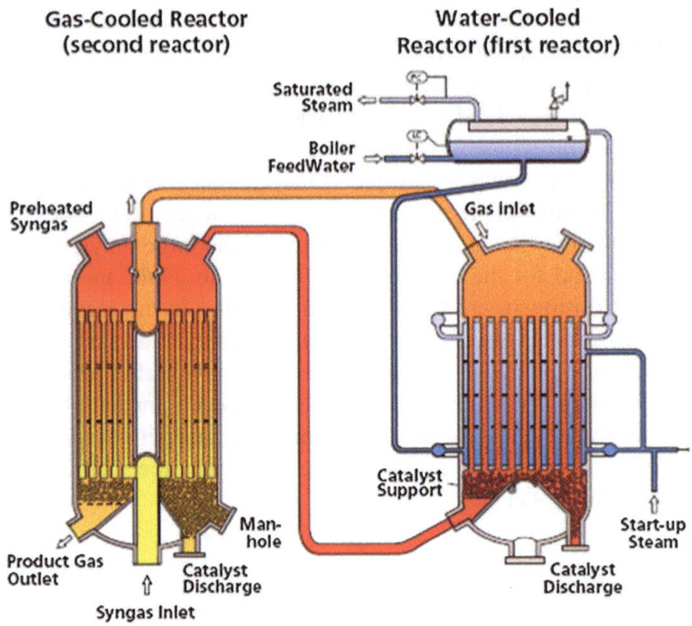

Fig. 5.5 Two-stage MegaMethanol® process offered by Lurgi [29]

water-cooled reactor and a gas-cooled reactor, see Fig. 5.5. The outlet gas from the water-cooled reactor serves as an inlet gas for the gas-cooled reactor. The temperature of the gas-cooled reactor is comparably low in order to achieve thermodynamically more favourable conditions. The temperature and space velocity in the water-cooled reactor are higher. As a result, the reactor size and catalyst bed volume can be reduced.

5.2.2.4 Liquid Phase Processes

The production of methanol using the liquid-phase synthesis process (LPMeOH technology) was developed by Air Products and Chemicals and demonstrated by Eastman Chemicals in Kingsport (Tennessee, USA) [30]. The LPMeOH process has not yet been realized on a commercial scale. This technology uses catalyst suspensions in mineral oil. Syngas flows through the liquid phase from the bottom to the top. Due to the intense gas-liquid mixing, this specific reactor concept guarantees beneficial conditions for the mass and heat transfer. The liquid phase synthesis is characterized by high operational flexibility. Syngas produced by coal gasification can be used without a prior CO shift process, because the LPMeOH process facilitates the direct utilization of fed gases with a H_2/CO ratio lower than [31]. However, one significant disadvantage compared to gas-phase methanol synthesis is the costly technical equipment, especially with regard to the catalyst preparation and the handling of the catalyst suspensions in the mineral oil.

5.2.3 *Production of Synthetic Liquid Fuels from Methanol*

Methanol can be used for the production of synthetic liquid fuels by one-stage or two-stage technologies. An overview of different processes is given in Fig. 5.6. Only one-stage technologies to convert methanol to synthetic liquid fuels will be described in this chapter. Using this approach, synthetic gasoline can be produced from methanol with high product selectivity. The consecutive reaction steps can be represented as follows [32]:

$$\begin{aligned} &2CH_3OH \leftrightarrow CH_3OCH_3 + H_2O \\ &CH_3OH, CH_3OCH_3 \rightarrow \text{Light Olefins}, H_2O \\ &\text{Light Olefins} \rightarrow C_{5+}\ \text{Olefins} \\ &C_{5+}\ \text{Olefins} \rightarrow \text{Paraffins, Naphthenes, Aromatics} \end{aligned} \tag{5.6}$$

The first reaction step is the dehydration of methanol under the formation of dimethyl ether (DME). DME can be used as a diesel substitute (see Fig. 5.2). A further dehydration step leads to the formation of light olefins. Finally, complex chemical reactions, including H-transfer, alkylation and polycondensation, produce a mixture of hydrocarbons, among them alkanes, olefins, aromatics and naphthenes. Virtually all oxygen is released with water; the organic phase contains only few oxygenates.

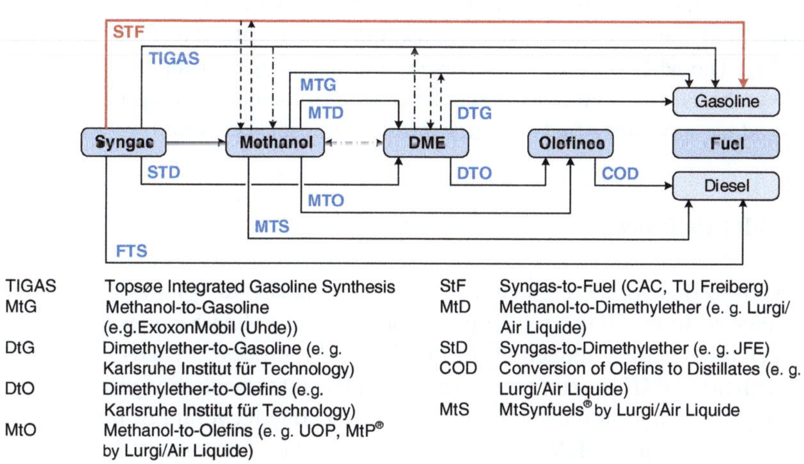

Fig. 5.6 Commercial technologies for different steps of the overall process chain from syngas to synthetic fuels

The typical product mixture of one-stage technologies is limited for compounds with up to 10 C atoms. This is caused by the specific reaction mechanism described in detail in Sect. 5.2.5. The high selectivity for the gasoline production is enabled by zeolites used as a catalyst. Zeolites are crystalline compounds (aluminosilicates) characterized by a highly ordered microporous lattice structure with uniform pore size distribution and pore geometry. The selectivity of the zeolite catalysts is caused by geometrical factors and is therefore referred to as "shape selectivity".

5.2.3.1 MtG Process

The technically most proven technology for the production of synthetic liquid fuels from methanol is the "Methanol to Gasoline®" process (MtG), developed in the 1970s by Mobil (now ExxonMobil) [33]. The MtG process was applied until 1997 in a plant in New Plymouth (New Zealand) with an annual capacity of 600 thousand tonnes of gasoline. The feed gas was produced from natural gas. Since 2009 a MtG plant with an annual capacity of 100 thousand tonnes of gasoline has been operated in China by Jincheng Anthracite Mining Group using syngas produced by coal gasification.

In the MtG process, methanol firstly undergoes a dehydration step to produce DME in an adiabatic fixed-bed reactor at 20–25 bar and an inlet temperature of 300–320 °C. The produced equilibrium mixture containing methanol, DME and water is supplied to the gasoline converter at an inlet temperature of 320–340 °C. The MtG process uses a zeolite catalyst of type ZSM-5. The reaction heat is used to generate high-pressure steam. The product mixture is separated into a gasoline-rich fraction, an aqueous fraction and gaseous products. Gases are recycled into the reactor. The gasoline fraction undergoes distillation and further post-treatment processes. Among other things, in alkylation reactions olefins are converted into iso-alkanes and the so-called "Heavy-Gasoline Treatment" (HGT) is to remove durene [34]. The MtG process possesses several advantages compared to Fischer–Tropsch synthesis. However, the direct conversion of methanol in liquid fuels is not able to produce any other fuels besides gasoline (diesel, kerosene).

5.2.3.2 TIGAS Process

Haldor Topsøe developed a process to convert syngas into synthetic gasoline; a process referred to as TIGAS™ (Topsøe Integrated Gasoline Synthesis) [35]. The main target was CAPEX reduction. In the TIGAS™ process, syngas is converted into a methanol/DME mixture and subsequently into synthetic gasoline. Both process steps have a joint recycle loop (without separate recycling of the unreacted syngas after the methanol/DME formation) and are therefore operated at the same process pressure of roughly 20 bar. The syngas can be directly converted into a methanol/DME mixture thanks to a multifunctional catalyst developed by Haldor Topsøe. The gasoline production step is similar to the MtG process by Mobil, with some minor modifications due to the implementation of a joint recycle loop for two process steps. The zeolite catalyst of the ZSM-5 type is also used.

5.2.3.3 STF Process

A further technology to convert syngas into gasoline, referred to as the "Syngas to Fuel" (STF) process, was developed by CAC GmbH and demonstrated in a pilot plant in Freiberg (Germany) [36]. It is distinguished from the competing commercial technologies by several benefits especially related to the higher product quality (lower content of aromatics, higher content of iso-alkanes) [37]. The better characteristics of the STF process are mainly down to the special construction of the isothermal reactors, enabling efficient heat recovery and therefore a more uniform temperature distribution inside the reactor. In addition, an improved catalyst with high selectivity for gasoline production is used.

5.2.4 Production of Olefins from Methanol

5.2.4.1 Market Analysis

Ethylene and propylene are important basic chemicals. Their worldwide production amounts to 140 million t/a of ethylene and 105 million t/a of propylene (in 2016) [38]. The share of methanol-based olefin production has steadily increased in recent years, especially in China. Methanol produced from coal-derived syngas is mostly used for conversion into olefins. Annual methanol production in China is nearly 30 million tonnes, approximately 10 million tonnes of which are converted into olefins [38]. Selected commercial plants for the methanol-based production of olefins are listed in Table 5.2.

Table 5.2 Selected plants for methanol-based olefin production

Plant	Country	Year	Technology	Ethylene, propylene (in kt/a)	Lit.
Shenhua, Baotou	China	2010	DMTO	600	[20]
Sinopec, Zhongyan	China	2010	S-MTO	200	[23]
Datang Intl., Duolin	China	2012	MTP™	460	[23]
Ningbo, Heyuan	China	2013	DMTO	600	[23]
Wison Clean Energy	China	2014	UOP/Hydro	300	[23]
Shenhua, Ningxia	China	2014	MTP™	500	[23]
Yilin Energy	China	2014	UOP/Hydro	600	[39]
Jiutai Energy, Ordos	China	2015	UOP/Hydro	600	[40]
Zhejiang, Jiaxin	China	2015	UOP/Hydro	600	[23]

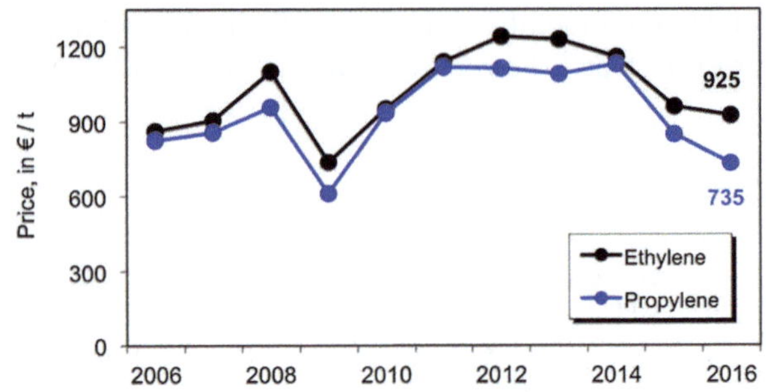

Fig. 5.7 Light olefin prices in Europe [41]

Prices for ethylene and propylene in Europe in the last 10 years are shown in Fig. 5.7. Ethylene and propylene are used for the production of various chemicals. The most important area of application is the production of polymers: polyethylene and polypropylene account for a market share of more than 50%.

5.2.4.2 Technology Description

In principle, the conversion of methanol into light olefins is described, like the MtG process, by Eq. 5.6 (see Sect. 5.2.3). However, methanol conversion terminates, thanks to the proper selection of reaction conditions, at the step of olefin formation [34].

Olefins are reactive intermediates on the pathway from methanol to synthetic gasoline. There are several possible ways to freeze the methanol conversion at the olefin formation stage [32]:

- *Reduction of the residence time and/or process temperature*
  The drawback in this case is the circumstance that the higher selectivity of the olefin production is associated with a lower degree of methanol conversion, so the unreacted methanol needs to be separated and recycled.
- *Reduction of the partial pressure of methanol*
  At pressures <1 bar both high selectivity for the olefin production and virtually complete methanol conversion can be achieved.
- *Adaptation of the SiO_2/Al_2O_3 ratio in the zeolite catalyst of ZSM-5 type*
  An increased SiO_2/Al_2O_3 ratio corresponds to a drop in the acidity of the catalyst and consequently to a decrease in the catalytic activity. Thus, the increase in the SiO_2/Al_2O_3 ratio in the zeolite catalyst ZSM-5 leads to an increased yield of olefins and a decreased yield of aromatics and C_1–C_4-alkanes (similar to the effect of reduced residence time).

Eventually, methanol is converted into light olefins at lower pressures (<3 bar) and slightly higher temperatures (450–500 °C) than in the case of the liquid fuel production [42]. Zeolites are used as a catalyst. The reaction mixture contains mainly gasoline and C_2–C_4 olefins. Appropriate technologies are offered by several companies.

5.2.4.3 MTO Process Developed by Mobil

The "Methanol to Olefins" (MTO) process was developed by Mobil to convert methanol into light olefins in a fluidized-bed reactor at temperatures of 470–515 °C and pressures of 2.2–3.5 bar [43]. An H-ZSM-5 type zeolite catalyst with an increased SiO_2/Al_2O_3 ratio is applied [44]. The reaction heat is removed by helical heat exchanger tubes placed inside the reactor under the generation of high-pressure steam [34]. The continuous catalyst regeneration enables the maintenance of constant catalyst activity. The mean catalytic activity is governed by the residence time of the catalyst in the regeneration unit. At the reactor outlet, the reaction mixture is separated by a distillation column into ethylene, light olefins and synthetic gasoline (which mainly contains aromatic compounds). The yield of light olefins varies in the range of 50–75%, depending on the reaction conditions of the MTO process [45].

5.2.4.4 MTP Process Developed by Lurgi

Lurgi developed the "Methanol to Propylene™" (MTP) process for the methanol-based production of light olefins. The process scheme is illustrated on Fig. 5.8. Methanol undergoes a catalytic dehydration step in a separate adiabatic fixed-bed reactor [46]. The reaction mixture produced, which contains methanol, DME and water, is mixed with the recycled olefin stream and fed to an adiabatic fixed-bed reactor. This reactor equipped with 5–6 catalyst beds is operated at 400–450 °C and 1.3–1.6 bar [47]. A commercial H-ZSM-5 type catalyst from Clariant is applied; this was specially developed for the MTP process [48]. This

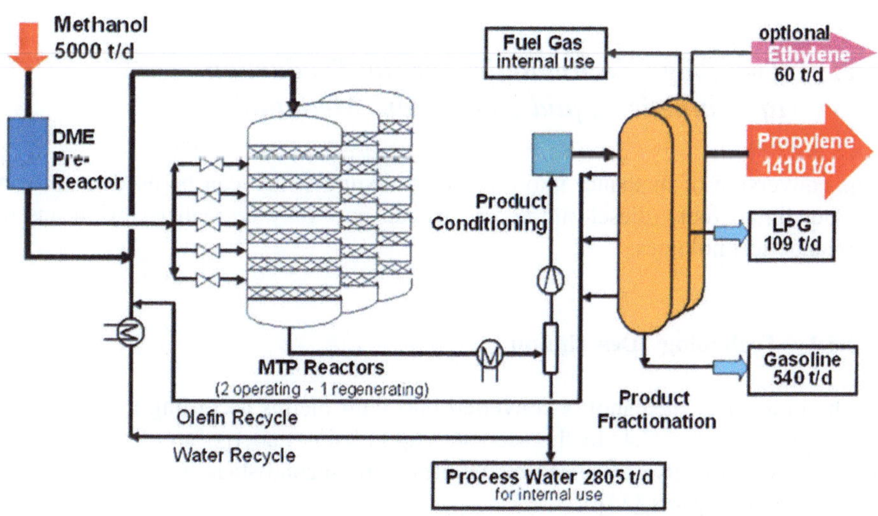

Fig. 5.8 Process scheme of the Lurgi MTP™ process [49]

catalyst is characterized by high selectivity with regard to propylene formation, a very low propane yield and a reduced tendency to coke deposition on the catalyst surface [48]. As a rule, the catalyst should be regenerated after roughly 500–600 operating hours [48].

The conversion degree for the methanol/DME mixture in the MTP process is higher than 99% [48]. The reaction product contains mainly propylene, gasoline and LPG, with a typical product ratio (on the mass basis) of propylene:gasoline: LPG = 13:5:1 (see Fig. 5.8).

5.2.4.5 MTO Process Developed by UOP/Hydro

A further MTO process to convert methanol into light olefins was developed by UOP (Des Plaines, Illinois) and Norsk Hydro (Norway). This technology is characterized by a higher selectivity for olefin formation compared to Mobil's MTO. The olefin yield can achieve up to 50% for ethylene and 30% for propylene [50, 51]. The reactor concept is similar to Mobil's MTO process [50]:

- a fluidized-bed reactor
- reaction heat is removed by heat exchangers, under generation of high-pressure steam

The essential difference is that the UOP/Hydro MTO process uses the SAPO-34 catalyst, which is characterized by very small pores [51]. This specific catalyst choice is presumably the reason for the higher formation of light olefins. The ethylene/propylene ratio can be changed under the influence of reaction conditions [50].

5.2.5 Two-Stage Technologies for the Production of Synthetic Liquid Fuels from Methanol

The conversion of methanol into a product mixture which can be used to recover heavier liquid fuels (diesel and kerosene) can now only be realized by means of two-stage technologies.

5.2.5.1 Technology Description

In the first stage, methanol is converted into light olefins according to procedures described in Sect. 5.2.4. In the second, stage olefins undergo an oligomerization process, which is also catalysed by zeolites. This second stage is characterized by the following aspects [34]:

- Oligomerization proceeds at lower temperatures (<300 °C) and higher pressures (>50 bar).
- The molar mass distribution of the product mixture is influenced by the reaction conditions: higher temperatures and lower pressures favour the formation of low-weight products.
- Iso-olefins are the preferential products of the oligomerization process, resulting in an improved quality of the produced synthetic liquid fuel.

Reaction mixture obtained after the oligomerization stage can be separated by distillation into different fuels (gasoline, diesel, kerosene).

5.2.5.2 MTO/MOGD Technology Developed by Mobil

Mobil developed a two-stage technology based on a combination of the MTO process (see Sect. 5.2.4) and the "Mobil Olefins-to-Gasoline/Distillate" (MOGD) process for olefin oligomerization. Light olefins are converted by means of the MOGD process into a mixture containing various unsaturated hydrocarbons with a medium molar mass. These compounds allow the separation of a diesel fraction. A diagram of the process used to combine the MTO and MOGD processes is shown in Fig. 5.9.

The product mixture obtained after the MTO process is separated into ethylene, light olefins and synthetic gasoline (which predominantly consists of aromatic hydrocarbons). Light olefins are mixed with a part of the produced gasoline and fed to the adiabatic MOGD reactor. The MOGD process is performed at 200–320 °C and 30–70 bar [52]. Zeolites of ZSM-5 type are used as a catalyst [52, 53]. The

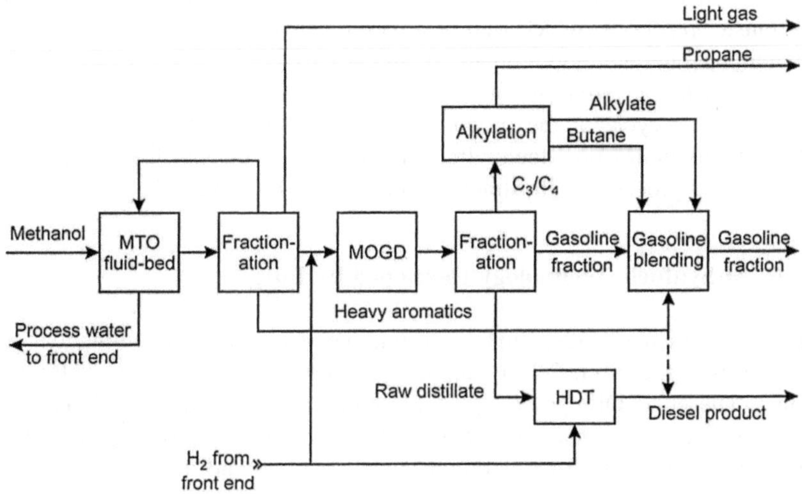

Fig. 5.9 Diagram of Mobil's MTO/MOGD technology process [43]

Table 5.3 Product distribution of the MOGD process in two operation modes (using C_3–C_4 olefins as feedstock), data from [53]

Fraction wt%	Gasoline mode	Distillate mode	
		Single pass	Recycle[a]
C_1–C_3	4	2	2
C_4	5	4	2
Gasoline (C_5 − 200 °C)	84	–	–
Heavy fraction (>200 °C)	7	–	–
Gasoline (C_5 − 165 °C)	–	29	12
Heavy fraction (>165 °C)	–	65	84

[a]With partial recycle of the produced gasoline

product mixture contains a high fraction of branched unsaturated hydrocarbons with 5–20 C atoms [45]. The chemical composition of the product mixture is virtually unrelated to the composition of the light olefin feedstock, since oligomerization and cracking reactions achieve an equilibrium state at the process temperature >230 °C: the product distribution is governed only by the temperature and pressure, provided that the residence time is long enough [52]. The MOGD process allows very flexible operation: the product ratio diesel/gasoline can be varied in a very broad range from 0 to 9 [53]. A typical product distribution in the gasoline mode and the distillate mode is given in Table 5.3.

The product ratio of diesel/gasoline is mainly governed by following process parameters:

- *Temperature and pressure of the MOGD process*
  At temperatures >300 °C and lower pressure, the formation of gasoline is thermodynamically favoured [53, 54]. This is caused by the dominating effect of H-transfer reactions, which lead to a partial conversion of olefins in a mixture containing saturated and aromatic hydrocarbons.
- *Recycle of the gasoline product*
  The partial recycle of the produced gasoline increases the yield of the final diesel/kerosene fraction due to the high reactivity of the gasoline components in the MOGD process, as illustrated by data given in Table 5.3.

5.2.5.3 MtSynfuels Technology Developed by Lurgi

Lurgi developed the "MtSynfuels®" technology based on a combination of the MTP process (see Sect. 5.2.4) and the "Conversion of low-molecular-weight Olefins to Diesel" (COD) process [46]. Light olefins produced by the MTP process are used as feedstock for the COD process, leading to the formation of a hydrocarbon mixture suitable for separation into a gasoline and a diesel fraction. The overall process is shown in Fig. 5.10.

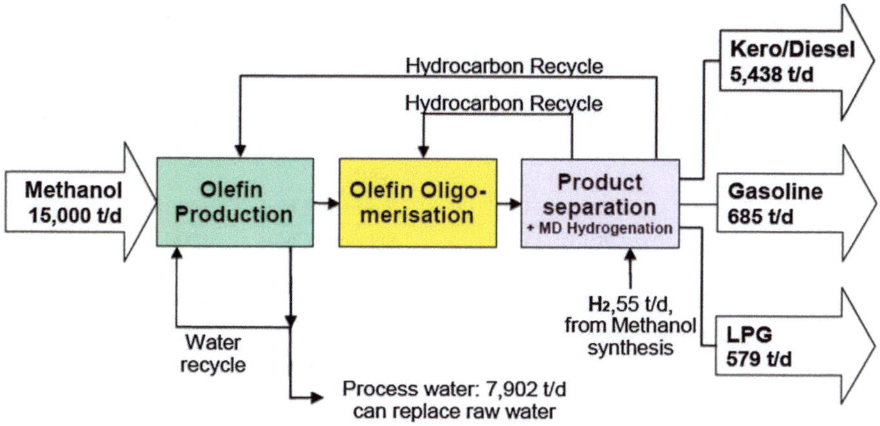

Fig. 5.10 Diagram of Lurgi's MtSynfuels® technology process [55]

The COD process was developed by Lurgi and commercialized by PetroSA in Mossel Bay (South Africa) in a large-scale plant with an annual capacity of 250 kt/a operated since 1993. The COD process is performed in adiabatic fixed-bed reactors at 200–280 °C and 45–55 bar [56, 57]. Zeolite COD-9 is used as a catalyst; it was developed by Clariant specially for the COD process [56, 58]. This catalyst can be applied not only for the oligomerization of light olefins, but also for the cracking and isomerization reactions of the heavy olefins produced.

5.3 Fischer–Tropsch Synthesis

Fischer–Tropsch synthesis, which enables the production of liquid fuels and chemicals, can be described in a simplified form by the Eq. 5.3 (see Sect. 5.1). This technology was developed in the 1930s and used in large-scale plants in Germany up to the end of the II World War and later in South Africa, starting from the 1950s. In recent years new Fischer–Tropsch plants were installed in the Middle East (with natural gas as feedstock for syngas production) and in China (with coal-derived syngas). Some commercial Fischer–Tropsch plants are listed in Table 5.4.

Low-temperature and high-temperature technologies are available for Fischer–Tropsch synthesis. High-temperature FT processes apply iron-based catalysts, low-temperature FT processes uses iron- and cobalt-based catalysts. Besides iron and cobalt as active components, catalysts contain further dopants, such as sodium potassium, manganese, vanadium, ruthenium, to specifically influence the catalytic activity and selectivity [64]. The temperature of FT synthesis does not exceed roughly 400 °C, due to the coke deposition at higher temperatures, which leads to a faster deactivation of the catalyst.

Table 5.4 Selected Fischer–Tropsch plants

Plant	Country	Year	Feedstock	Technology	FT product (in kt/a)	Lit.
Sasol I	S. Africa	1955	Coal	ARGE	300	[59]
Sasol II/III	S. Africa	1981	Coal	Sasol advanced synthol	7200	[60]
Yitai	China	2012	Coal	Synfuels China	1000	[61]
Shanxi Luan	China	2015	Coal	Synfuels China	1000	[61]
Shenhua	China	2016	Coal	Synfuels China	4000	[61]
Mossgas	S. Africa	1993	NG	Synthol	2000	[60]
Bintulu	Malaysia	1993	NG	Shell middle distillate	660	[59]
Pearl	Qatar	2011	NG	Shell middle distillate	6300	[62]
Oryx	Qatar	2006	NG	Slurry-bed	1530	[60]
Escravos	Nigeria	2014	NG	Slurry-bed	1530	[63]

One typical aspect of FT synthesis is the formation of a very broad product spectrum containing various saturated, unsaturated and aromatic hydrocarbons in a large molar mass range. The product distribution depends on the process temperature. Lower temperatures favour the formation of long-chain hydrocarbons, while more short-chain products are formed at higher temperatures. The specific product distribution depending on the molar mass is governed by the Schulz-Flory function. A typical example of the relationship between product formation and chain growth probability α is given in Fig. 5.11. The catalyst choice also influences the product spectrum. Depending on the applied catalyst and other process conditions, the predominant product of FT synthesis may be gasoline (e.g. in the Synthol process),

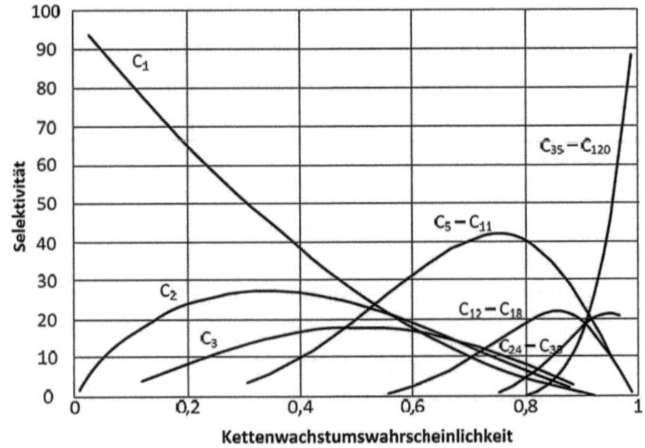

Fig. 5.11 Typical product distribution for Fischer–Tropsch synthesis [65]

diesel (e.g. in the ARGE process), light olefins (e.g. in the Dow LPG process) or FT wax (e.g. in the Shell Middle Distillate process). The maximum product yield is 50% for gasoline and nearly 40% for diesel.

5.3.1 Low-Temperature Fischer–Tropsch Synthesis

Low-temperature FT synthesis processes are characterized by the preferential formation of products with a higher molar mass. These technologies are suitable for the production of diesel and waxes (lubricants). The major drawback is the need for a complex post-treatment process to upgrade the FT raw product, which includes several process steps and requires costly facilities. Low-temperature FT technologies are offered by several vendors; the most well-proven technologies from a technical viewpoint are the ARGE process and the Slurry-Bed process, see Fig. 5.12.

The ARGE process applies isothermal tubular reactors with slim tubes filled with catalyst. FT synthesis proceeds at 220–250 °C, 20–30 bar with a $Fe/Cu/K_2O/SiO_2$ catalyst. The tubes are flushed with boiling water to recover the released reaction heat. The obtained high-molecular product (FT wax) flows down to the reactor bottom and is discharged through a wax separator. The fraction of FT waxes in the product mixture is at least 50%, and the output of low-weight products (gasoline fraction) is much lower.

Slurry-Bed process uses catalysts suspended in a liquid phase. Syngas flows through the liquid phase from the bottom to the top, while the FT wax produced is

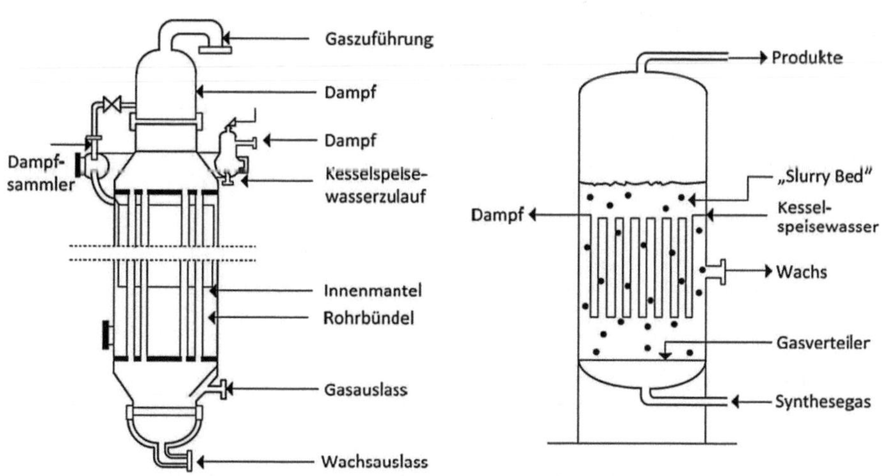

Fig. 5.12 ARGE reactor (left) and Slurry-Bed reactor (right) [66]

discharged at the bottom of the reactor. Reaction heat is removed using heat exchangers immersed in the liquid phase. The Slurry-bed process is characterized by a higher product yield and quality compared to the ARGE process, thanks to the more uniform temperature distribution inside the reactor. Further benefits are lower CAPEX and higher flexibility.

5.3.2 *High-Temperature Fischer–Tropsch Synthesis*

High-temperature FT synthesis processes are applied for the preferential production of gasoline and light hydrocarbons. All technologies available use fluidized-bed reactors. The former Synthol process has been replaced by the more efficient Sasol-Advanced-Synthol (SAS) process. Both reactor concepts are depicted in Fig. 5.13.

The Synthol process utilizes a low-cost iron catalyst obtained by the reduction of magnetite. FT synthesis is performed at 300–340 °C and 20–25 bar. The product mixture, containing unconverted syngas, reaction products and a catalyst, is fed to a cyclone in order to separate the catalyst particle from the stream.

The improved Sasol-Advanced-Synthol (SAS) process is operated at higher pressures up to 40 bar, using a special $Fe/Cu/K_2CO_3$ catalyst. This catalyst is characterized by a lower tendency to deactivation. The achieved degree of conversion is significantly higher than in the case of the Synthol process, due to the higher activity of the applied catalyst. An important benefit of the SAS process is the less abrasive impact on the inner reactor walls, thanks to the lower gas velocities needed for this technology.

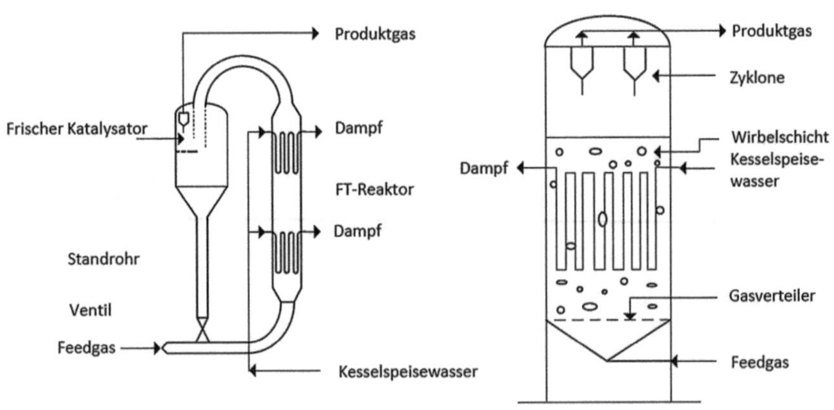

Fig. 5.13 Synthol reactor (left) and Sasol-Advanced-Synthol reactor (right) [66]

5.4 Synthetic Natural Gas

5.4.1 Market Analysis

Coal gasification with the subsequent conversion of syngas into methane (SNG—Synthetic Natural Gas) is a technically well-proven, commercially realized technology. But it should be noted that the production costs for SNG cannot compete with the natural gas price, due to the high CAPEX for syngas generation and gas cleaning facilities. For this reason, there are only few commercial applications for coal-to-SNG worldwide (see Table 5.5). Most SNG plants are currently located in China. The driving force here is the aspired economic independence from natural gas imports.

5.4.2 Technology Description

The conversion of syngas to SNG proceeds in line with Eq. 5.4 (see Sect. 5.1). As in the case of the methanol synthesis and FT synthesis, SNG production is described by a chemical equilibrium, which favours the formation of methane by increased pressure and decreased temperature. SNG synthesis is performed at 50–80 bar, 250–650 °C with nickel-based catalysts [70]. Due to the highly exothermic reaction heat, an efficient concept to recover reaction heat from the

Table 5.5 Selected Coal-to-SNG plants

Plant	Country	Year	Technology	SNG (million m^3/a)	Lit.
Great Plains Synfuels	USA	1984	Lurgi	1700	[67]
Datang Power, Hexingten	China	2012	Lurgi	4000	[61]
Datang Power, Chifeng	China	2013	Johnson Matthey/ Davy	4000	[68]
Qinghua Group, Yili City	China	2013	TREMP™	1400	[61]
Huineng Coal Power	China	2014	TREMP™	1600	[61]
CPI Xinjiang Energy	China	2015	TREMP™	6000	[68]
POSCO, Gwangyang	S. Korea	2016	TREMP™	700	[69]
Xinwen Energy, Yili City	China	2016	Johnson Matthey/ Davy	4000	[68]
Datang Power, Fuxin	China	2016	Lurgi	4000	[61]

methane converter plays a dominant role. Technologies based on fixed-bed reactors and fluidized-bed reactors have been developed for SNG production. Usually, several adiabatic fixed-bed reactors connected in series are used to achieve efficient heat recovery; the product mixture is cooled down at the outlet of each reactor. In the case of fluidized-bed technologies, a single reactor is enough to ensure efficient reaction heat recovery. But fluidized-bed reactors are disadvantaged by relatively high wear to the catalyst pellets.

The most widespread methanation technology is the Lurgi process, which is based on the application of two fixed-bed reactors. Among other places, this technology is implemented in the Great Plains Synfuels Plant in North Dakota (USA), which has been in operation since 1984. Haldor Topsøe has developed the TREMP™ process (Topsøe Recycle Energy-efficient Methanation Process) using three fixed-bed reactors. A diagram of the TREMP™ process is given in Fig. 5.14.

Further fixed-bed technologies have been developed by Johnson Matthey/Davy (HICOM process) [71] and Clariant/Foster Wheeler (Vesta process) [70]. Examples of fluidized-bed technologies are the Synthane process and the Bi-Gas process [71], but neither technology has been demonstrated on a commercial scale.

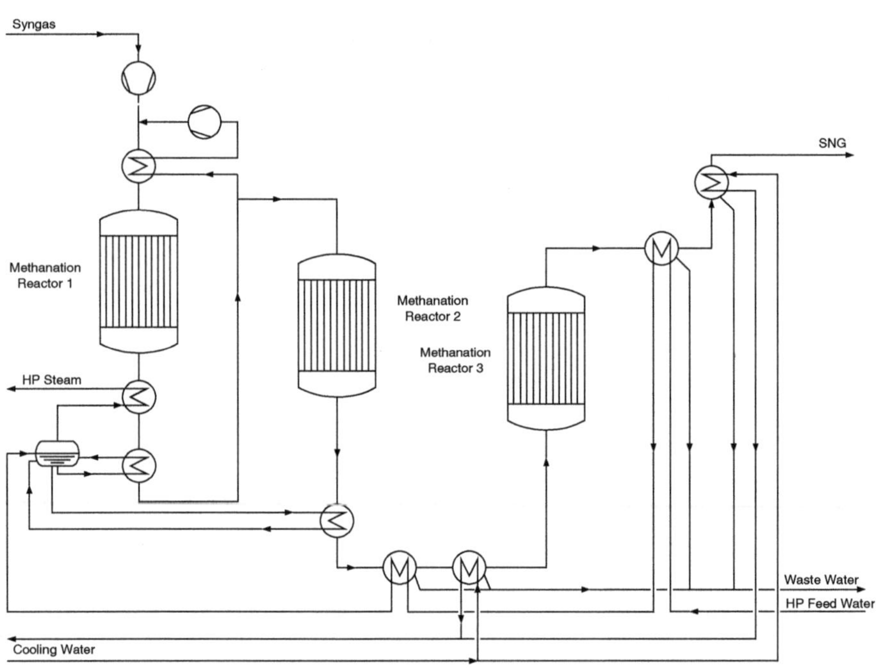

Fig. 5.14 Process scheme of the TREMP™ process by Haldor Topsøe [70]

5.5 Ammonia

5.5.1 Market Analysis

Worldwide ammonia production is roughly 200 million t/a (in 2016) [72]. The contribution of the coal-based route to ammonia production has increased greatly in the last decade. Currently, up to 30% of the ammonia produced results from syngas obtained by coal gasification. Most coal-based ammonia plants are located in China. The annual ammonia production in China amounts to 65 million t/a, nearly 80% of which is produced from coal [73]. Some of the coal-based ammonia plants are listed in Table 5.6. Approximately 85% of the ammonia produced is used as feedstock for the fertilizer industry (urea, ammonia sulphate, ammonia nitrate, ammonia phosphate) [74]. In addition, ammonia is used as a feedstock for the chemical industry, to produce nitric acid, urea and various secondary chemicals (polyamide, polyurethane, polyacrylonitrile, resins). The market price for ammonia in Europe fluctuated in 2016 in a range between 260 and 320 €/t [75].

5.5.2 Technology Description

Ammonia synthesis is a technically well-proven technology which has been implemented worldwide on an industrial scale for over more than 100 years with no fundamental modifications. Ammonia is synthesized at 400–500 °C and 150–250 bar using iron-based catalysts [74]. The reaction is described by a chemical equilibrium, which favours the formation of ammonia by increased pressure and decreased temperature. The degree of conversion after a single reactor pass is usually about 25–35%. The unconverted H_2/N_2 mixture is separated from the ammonia product by means of cryogenic condensation at temperatures below 0 °C and recycled back to the ammonia converter.

Table 5.6 Selected plants for the coal-based ammonia production

Plant	Country	Year	Technology	Ammonia (in kt/a)	Lit.
Anhui Huainan	China	2009	Casale	480	[76]
Bodashidi, Wushen	China	2014	Kellogg-Brown-Root	530	[77]
Hegang Huahe Coal	China	2014	Haldor Topsøe	350	[77]
Henan Jinmei	China	2014	Casale	350	[77]
Zhongyuan Kunshan	China	2014	Haldor Topsøe	350	[77]
China Coal, Ordos	China	2015	Kellogg-Brown-Root	1000	[77]
Henan Junhua	China	2015	Casale	650	[77]
Wulan Coal, Ordos	China	2015	Haldor Topsøe	750	[77]
Yangmei Taiyan Chem.	China	2015	Casale	350	[77]
Xing'an Berun Chem.	China	2015	Casale	350	[77]

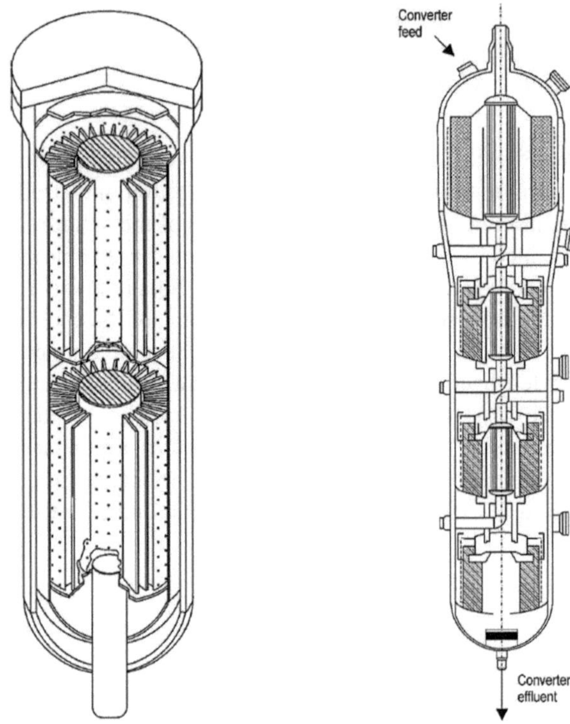

Fig. 5.15 Isothermal ammonia converter by Casale (left) and adiabatic Kellogg converter (right) [74]

Various fixed-bed technologies are available for ammonia synthesis. As in the case of other routes for syngas utilization, reaction heat recovery is an important aspect in process design. For this reason, quasi-isothermal tube reactors or adiabatic reactors with several catalyst beds are applied. Two examples of the reactor design are given in Fig. 5.15. In general, tube reactors contain tubular heat exchangers, which are placed inside the catalyst bed. Cold-fed gas flows through the tube to remove the reaction heat and immediately preheat the syngas to the reactor inlet temperature. Licensors for such technologies include ICI, NEC or Casale. Most of the commercial ammonia plants operated today use adiabatic reactors equipped with several catalyst beds. They are characterized by an easier and cheaper reactor design and at the same time by efficient heat recovery. Usually, cold-fed gas is inserted between the catalyst beds for cooling purposes. Especially in the case of ammonia plants with a higher annual capacity, special heat exchanger tubes are used for heat recovery, because this reactor design is characterized by increased thermal efficiency. Licensors include Uhde, BASF, Casale, Kellogg or Haldor Topsøe.

5.6 Selection of Syngas Utilization Routes

Considering technological and economic aspects, the following two routes to syngas-based products are chosen to analyse and assess complete process chains for non-fuel coal utilization in Russia:

- methanol production from syngas
- conversion of methanol into synthetic gasoline

Methanol is one of the most important basic feedstocks for the chemical industry, with annual worldwide production forecasted to increase at 6% per year [16]. Most methanol production is based on syngas utilization. Coal gasification plays a constantly increasing role in this context: the share of methanol produced from coal-based syngas has already reached 40%. Depending on the local boundary conditions and the national economic and resource strategy, methanol could definitely be a feasible and competitive product for non-fuel coal utilization. The world methanol price has varied in the last decade in the range of 250–500 €/t. However, the fact should be taken into account that the discontinuous world economic development more strongly impacts the methanol price than, for example, the price of methanol-derived products, such as ethylene or propylene.

There are also some secondary products which can be efficiently synthesized from methanol. As an example, this study considers the conversion of methanol into ethylene and propylene according to the "Methanol to Olefin" process. Both these compounds are also important feedstock, especially for polymer production. The market demand is increasing continuously, with the annual worldwide production forecasted to grow by 4–5% and 5–6% per year for ethylene and propylene, respectively [78]. However, these chemicals were not taken into consideration as feasible products of syngas utilization under the general conditions of the Russian coal industry. The reason is that this option for non-fuel coal utilization presupposes the existence of a large and highly integrated chemical park to ensure that the gaseous intermediates ethylene and propylene can be further processed near the syngas generation facility. Usually, this is the case in regions with a powerful chemical industry. An alternative option which would be especially beneficial for the Kuzbass coal region could be the production of methanol for the Chinese market. At the moment, the rate of increase in the production capacities of MTO plants in China is higher than the corresponding growth of coal gasification facilities. Due to this mismatch, some of the newly installed MTO capacities in China are now already envisaging purchasing methanol on the world market.

A further option which was considered in this study is the conversion of methanol into synthetic liquid fuels. Methanol can be used for the production of synthetic liquid fuels by one-stage (e.g. MtG) or two-stage (e.g. MtSynfuels®) technologies. The benefit of the two-stage technologies is the general possibility to produce various fuels (gasoline, diesel, kerosene), but they are disadvantaged by the need for higher CAPEX. Using one-stage approaches, methanol can only be converted into synthetic gasoline. But these technologies require less complex facilities and are therefore economically more feasible. This study selects the conversion of methanol into gasoline using the MtG process to analyse whole process chains, from the coal to the final product.

An interesting option for syngas utilization that fits Russian coals could be ammonia production. This conversion route is not yet taken into consideration since ammonia, as a coal-derived product, also presupposes the existence of the fertilizer industry near the coal gasification plant.

Chapter 6
Selected Processes

Based on the detailed analysis described in Chaps. 2 (Investigated Coals), 3 (Gasification Technologies) and 5 (Technologies of Syngas Utilization), two gasification technologies and two syngas conversion routes are chosen as the most feasible ones that fits the Russian coal industry. The most promising gasification technologies are identified as the Siemens entrained-flow gasification and the new COORVED staged gasification, comprising a fixed bed at the bottom and a jetting fluidized bed above. Among the syngas utilization products, methanol and synthetic gasoline were selected. An overview of the overall process chains and modelled cases considered is given in Fig. 6.1.

Individual processes within the overall process chains considered are described in detail in this chapter. In addition, the basic boundary conditions for process modelling using the commercial software Aspen Plus® are given.

6.1 Gasifier

Both gasification technologies considered use raw coal, crushed to the required size range. The COORVED gasifier utilizes a broad coal fraction in the range of 0–6 mm. Entrained-flow gasification requires coal fines <250 µm. The Siemens gasifier uses dense coal fed with CO_2 provided as an off-gas from the acid gas removal unit. The COORVED gasifier works with a gravitational solids feeding system.

Coal gasification requires oxygen, which is supplied by the air separation unit (ASU). Usually, cryogenic air liquefaction and separation is applied. The flow chart of the ASU is given in Fig. 6.2.

An oxygen purity of 99 vol.% was presupposed for the model. The high oxygen purity is required in order to limit the concentration of inert components in the

V. Litvinenko and B. Meyer, *Syngas Production: Status and Potential for Implementation in Russian Industry*,
https://doi.org/10.1007/978-3-319-70963-5_6

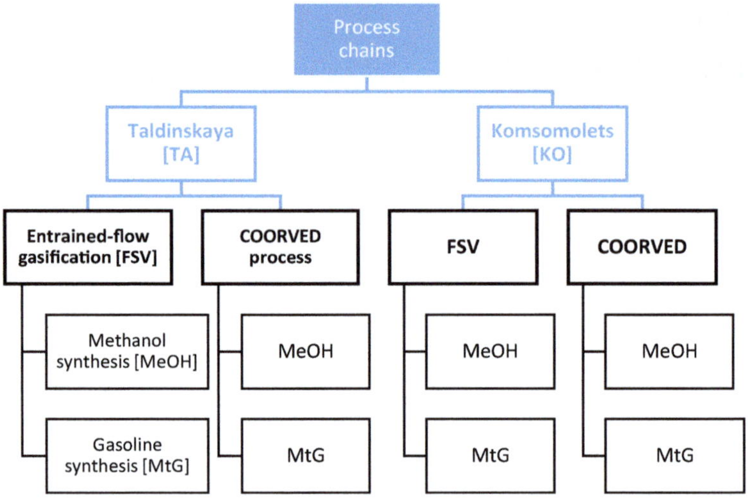

Fig. 6.1 Overview of the processes and calculation cases considered. *Note* COORVED = CO_2-Reduktion durch innovatives Vergaserdesign

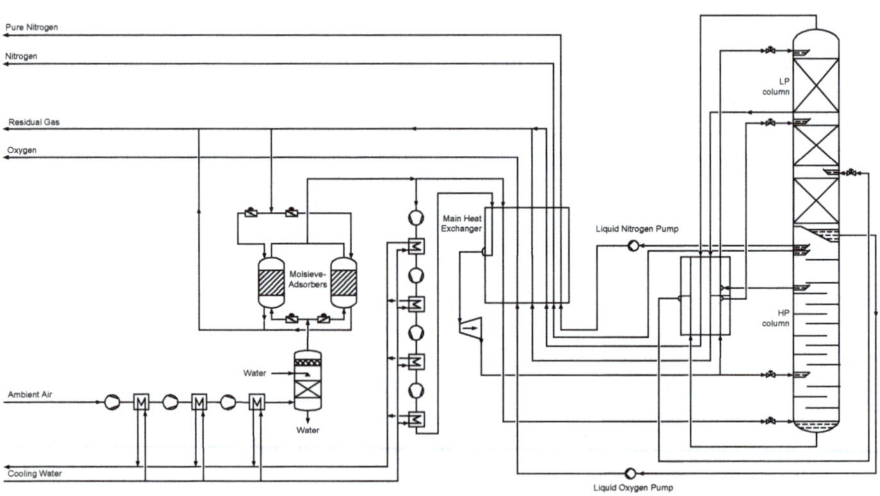

Fig. 6.2 Flow chart of the cryogenic air separation unit

syngas produced. In the case of the synthesis routes considered which involve a recycle loop for unreacted syngas, significantly higher concentrations would lead to inert components accumulating in the recycle gas and therefore to an increased energy demand for the syngas compressors and higher purge gas splits. Further boundary conditions for modelling are given in Table 6.1.

Table 6.1 Basic boundary conditions for modelling the low-pressure air separation unit

Parameter	Unit	Value
Outlet temperature after cooling of the compressed air	°C	30
Outlet temperature evaporation cooler	°C	20
Outlet pressure after the decompression of the liquid air	bar	14
Outlet temperature product gases (GOX, GAN)	°C	$t_{out} \approx t_{ambient}$
Pressure at the top of the high-pressure column	bar	5.12
Pressure at the top of the low-pressure column	bar	1.31
Pressure drop in the high-pressure column	mbar	175
Pressure drop in the low-pressure column	mbar	40
Theoretical stages in the high-pressure column	–	35
Theoretical stages in the low-pressure column	–	25

6.1.1 Entrained-Flow Gasifier

The entrained-flow gasifier developed by Siemens is equipped with an integrated cooling screen inside the gasifier, which is used to generate saturated intermediate-pressure steam. The syngas produced is directed with a high gas velocity (approx. 20 m/s) from the gasifier into the water quench chamber placed directly below the gasifier. Preheated water is sprayed into the hot gas stream to suddenly cool the gas down to the water steam dew point. From commercial gasifier operation it is known that the high partial pressure of the water steam in the water quench chamber enhances the homogeneous water-gas shift reaction. In this context, an additional CO conversion of 3% compared to the syngas composition at the reactor outlet was accounted for in the gasifier modelling. In the same water quench, liquid slag drops into the water bath and solidifies. The granulated slag is discharged at the bottom of the quench chamber. Water-saturated raw gas leaves the quench chamber at a temperature of 200–250 °C. The Siemens gasifier is shown schematically in Fig. 3.3 (see Sect. 3.3.3).

When modelling the Siemens gasifier, the syngas temperature at the reactor outlet was fixed 150 K higher than the ash fluid temperature of the respective coal, in order to ensure that ash fusion was complete. For this reason, different gasifier temperatures were implemented for modelling in the case of the TA and KO coals. Further boundary conditions are summarized in Table 6.2.

6.1.2 COORVED Gasifier

The COORVED gasifier is described in detail in Sect. 3.3.4. For the modelling, a virtually complete carbon conversion was presupposed. Therefore, the ash released from the COORVED gasifier is carbon-free and can be dumped without costly post-combustion, as in the case of conventional fluidized-bed gasification, e.g.

Table 6.2 Basic boundary conditions for modelling the Siemens entrained-flow gasifier

Parameter	Unit	Value
Gasifier pressure	bar	40
Gasifier temperature, Taldinskaya coal	°C	1650
Gasifier temperature, Komsomolets coal	°C	1550
Coal loading of the CO_2 stream for the dense coal fed	kg/m^3	350
Gasification agent temperature at the reactor inlet	°C	240
Steam pressure at the outlet of the cooling screen	bar	47

Table 6.3 Basic boundary conditions for modelling the COORVED gasifier

Parameter	Unit	Value
Gasifier pressure	bar	40
Syngas temperature ex freeboard	°C	1200
Syngas temperature at the reactor outlet	°C	950
Pressure of the high-pressure steam in the convective heat exchanger	bar	140
Pressure of the intermediate-pressure steam in the convective heat exchanger	bar	40

HTW. Hot syngas is partially quenched at the top of the COORVED reactor, just before the inlet into the convective gas cooler. The sensible heat of the hot syngas is utilized to generate high-pressure and intermediate-pressure steam. The design of the CORVED gasifier is schematically represented in Fig. 3.4 (Sect. 3.3.4).

Due to the use of O_2/H_2O mixtures with high O_2 concentrations as the primary gasification agent, local temperatures in the hot zone exceed the ash softening temperatures (TA: 1440 °C, KO: 1310 °C), resulting in partial char agglomeration. Char agglomerates move downwards and fall on the fixed bed at the bottom of the COORVED gasifier. The secondary gasification agent is a mixture of oxygen and steam, with a steam concentration of 7 kg steam/m^3 (STP) oxygen, and is introduced via a rotating grate placed at the bottom of the gasifier. It ensures that the carbon conversion is complete in the bottom ash. Basic boundary conditions for the modelling are given in Table 6.3. The syngas temperature at the freeboard outlet is fixed for both coals to 1200 °C. The gas temperature is decreased to 950 °C by partial water quenching before the inlet into the gas cooler.

6.2 Gas Cleaning and Gas Treatment

The cooled syngas is fed to the gas cleaning system, which consists of three units: pressurized water scrubbing, sour CO shift and two-stage acid gas removal. The separated acid gas is utilized in a Claus plant for sulphur production.

Table 6.4 Basic boundary conditions for modelling the water scrubbing

Parameter	Unit	Value
Stage number	–	4
Inlet pressure	bar	38
Pressure drop	bar	1

6.2.1 *Pressurized Water Scrubbing*

Water scrubbing is required to separate the residual flue ash and remove ammonia and chlorides. This process was described using a simple absorber column model with 4 stages. The applied model accounts for dissociation reactions in the aqueous phase. Most of the washing water is recycled. Approximately 10% of the washing water is removed in order to avoid an accumulation of harmful substances (for example, the maximal allowed concentration of chlorides is 7000 ppm). Basic boundary conditions for the modelling are summarized in Table 6.4.

6.2.2 *Sour Gas Shift*

After the water scrubbing, the required syngas composition (stoichiometric number) is adjusted using the sour gas shift process. Part of the syngas stream is fed to the shift reactors, while the residual syngas is bypassed. Usually, two adiabatic fixed-bed reactors are used, with cobalt molybdenum sulphides as a catalyst. This catalyst is insensitive to sulphur impurities in the raw gas. The required steam concentration for the water-gas shift reaction is attained by adding intermediate-pressure steam. This is necessary only in the case of the relatively dry, unsaturated syngas produced by the COORVED gasifier. The exothermic heat of reaction is removed by steam generation. The temperature at the reactor inlet is 280 °C, and the maximum temperature at the reactor outlet is limited to 490 °C in order to avoid catalyst degradation. The applied catalyst simultaneously supports the COS hydrolysis reaction. Thus, no additional hydrolysis reactor is required before the acid gas removal. A flow chart of the sour gas shift process, together with the water scrubbing, is given in Fig. 6.3. Basic boundary conditions for the modelling are given in Table 6.5.

6.2.3 *Acid Gas Removal*

The product gas after the CO shift process is cooled down and the water condensate is discharged. In the next step, syngas is fed to the acid gas removal (AGR) unit in order to remove CO_2, sulphur compounds and the remaining NH_3 and chlorides. The AGR process was modelled as selective, two-stage scrubbing (related to the

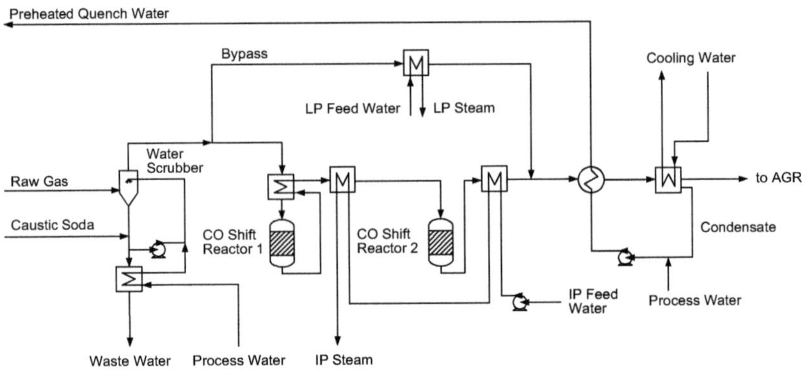

Fig. 6.3 Flow chart of the water scrubbing with subsequent two-stage sour gas shift

Table 6.5 Basic boundary conditions for modelling the sour gas shift

Parameter	Unit	Value
Steam/Dry gas ratio	mole/mole	>1.5
H_2S/H_2O ratio	mole/mole	<0.001
Reactor inlet temperature	°C	280
Reactor outlet temperature (first reactor)	°C	<490

Rectisol process) with methanol as a physical solvent. Compared to the competing scrubbing processes (e.g. Selexol), physical scrubbing with methanol is characterized by a higher syngas purity and high H_2S/CO_2 selectivity. The target value for the H_2S concentration in the purified syngas is fixed for all calculation cases at 0.1 ppmv, in order to meet the gas purity requirements for methanol synthesis (see Table 6.6).

The AGR process is implemented at a temperature of −50 °C, using an ammonia chiller for cooling purposes. To minimize the relatively high energy demand for the

Table 6.6 Gas quality specifications for methanol synthesis

Component	Unit	Value
CO	vol.%	30–32
H_2	vol.%	64–68
CO_2	vol.%	Compatible
$CH_4 + N_2 + Ar$	vol.%	<4
Ar	vol.%	No
H_2O	–	Inert
Sulphur (overall)	mg/m^3 (STP)	<0.1
Alkaline	mg/m^3 (STP)	<0.2
Dust	mg/m^3 (STP)	<0.2

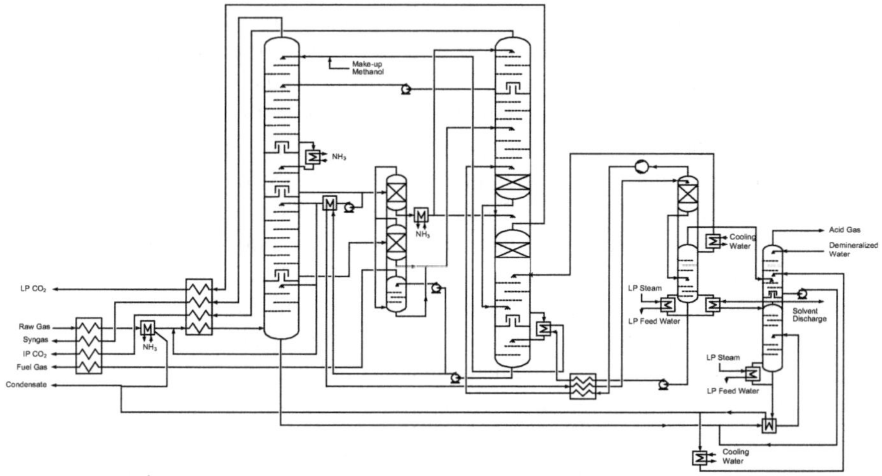

Fig. 6.4 Flow chart of the selective acid gas removal

Table 6.7 Basic boundary conditions for modelling the acid gas removal

Parameter	Unit	Value
Scrubbing temperature	°C	−40
CO_2 content in the purified syngas	vol.%	<3.0
H_2S content in the purified syngas	ppmv	<0.1
Plate number for the H_2S absorber		5
Plate number for the CO_2 absorber		18
Temperature at the top/bottom of the regeneration column	°C	30/95

chiller operation, heat integration within the overall plant was taken into consideration. Gas scrubbing proceeds in two absorber columns, the first for H_2S removal, the second for CO_2 removal. Cold methanol, charged with H_2S and CO_2, respectively, is fed to the regeneration unit, including the regeneration by decompression and the thermal regeneration. The desorbed off-gas a with high H_2S content is sent to the Claus plant. The flow chart of the modelled acid gas removal scheme is represented in Fig. 6.4. Basic boundary conditions are listed in Table 6.7.

6.2.4 *Claus Plant*

The produced H_2S-rich off-gas from the acid gas removal unit is further treated in a Claus plant, in order to convert H_2S into elemental solid sulphur. Sulphur is sold as a by-product; alternatively, the production of sulphuric acid, which is an important feedstock for the chemical industry, could be taken into consideration. Several plant configurations are available for the Claus process, which differ with respect to some basic features (oxidation agent, tail gas treatment). The Claus plant type which was chosen for modelling uses oxygen-enriched air as an oxidation agent. The tail gas is recycled to the sour syngas stream upstream of the AGR unit.

In the thermal stage of the Claus process, all the H_2S-rich off-gas is fed to the combustion chamber, where the off-gas is burned under oxygen deficiency conditions (one third of the full oxygen demand). The throughput is increased by the use of oxygen-enriched air, which requires a special burner construction. Some elemental sulphur (S_2 or S_8) is already produced in the combustion chamber. The heat of reaction is utilized via a convective heat exchanger placed at the outlet of the combustion chamber. The intermediate-pressure and low-pressure steam produced is used at different sites of the overall plant. Sulphur is condensed as S_8 in water-cooled drums at approx. 70 °C and discharged. The remaining gas is treated in two so-called contact beds (simple adiabatic fixed-bed reactors), with intercooling after each reactor and sulphur condensation in further water-cooled drums. The off-gas after the catalytic treatment contains some trace compounds (COS, CS_2) and therefore undergoes a hydrolysis step prior to recompression and recycling to the acid gas stream. The flow chart is shown in Fig. 6.5. Table 6.8 lists the boundary conditions of the modelling.

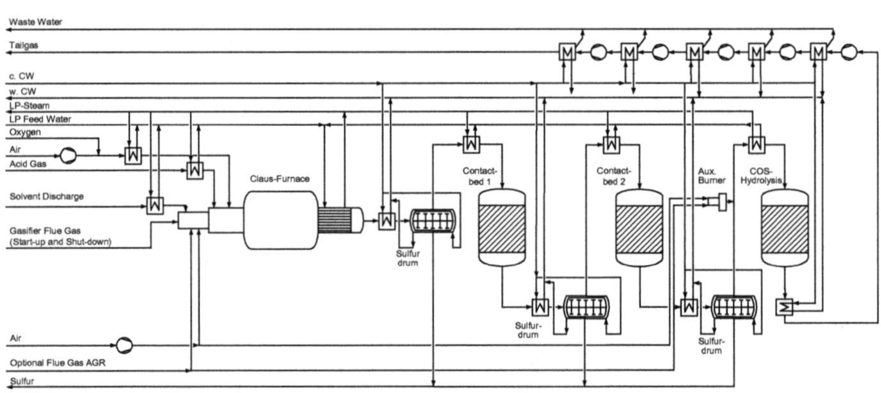

Fig. 6.5 Flow chart of the Claus plant

Table 6.8 Basic boundary conditions for modelling the Claus process

Parameter	Unit	Value
O_2 concentration of enriched combustion air	vol.%	40
Temperature at the inlet to the catalyst bed	°C	200
Temperature at the inlet to the COS hydrolysis unit	°C	150

6.3 Methanol Production

After the gas purification and conditioning, syngas is compressed and supplied to the methanol synthesis unit. Adiabatic and quasi-isothermal reactors are commercially available for methanol synthesis, which can be realized as a gas phase process in fixed-bed reactors or as a liquid phase process. The corresponding technologies were described in detail in Sect. 5.2.2. In this study, the methanol production was modelled following the quasi-isothermal Lurgi process. Due to the comparably small plant capacity chosen for economic considerations (see Sect. 7.1), a single-stage methanol synthesis was modelled. Methanol plants with higher annual generally use two-stage synthesis technologies, e.g. Lurgi's MegaMethanol® process (see Sect. 5.2.2).

The quasi-isothermal tubular reactor developed by Lurgi contains several hundred tubes, which are cooled with boiling water to remove the exothermic heat of reaction for the production of saturated intermediate-pressure steam. Therefore, the Lurgi reactor can be operated in a nearly isothermal mode at optimal temperatures of 240–260 °C. The selectivity of the methanol formation usually exceeds 99%. The stoichiometric number of the syngas at the inlet of the methanol reactor should be 2.05–2.10 and the CO_2 concentration should be at least 2.5 vol.%. The unreacted syngas is recycled to the methanol reactor, where part of it is removed as a purge gas to stop the inert components from accumulating in the recycle loop. The flow chart of the modelled methanol synthesis is shown in Fig. 6.6. Basic boundary conditions taken into account for modelling the methanol synthesis are listed in Table 6.9.

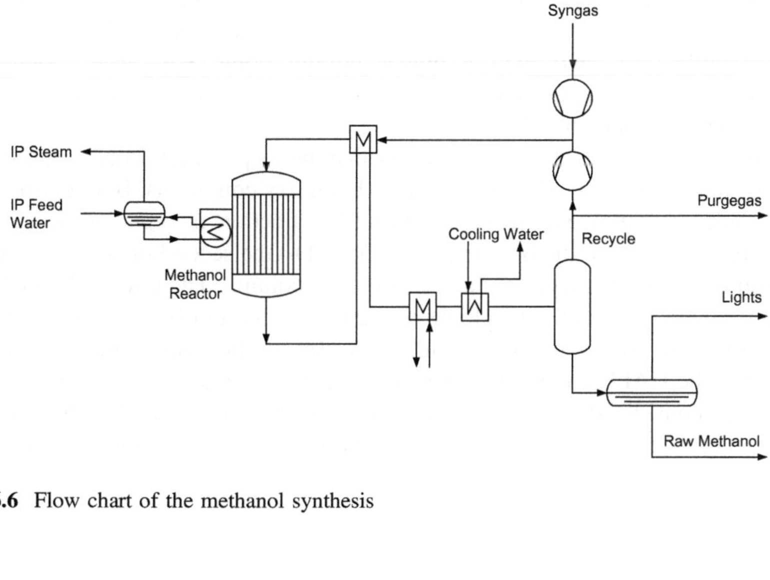

Fig. 6.6 Flow chart of the methanol synthesis

Table 6.9 Basic boundary conditions for modelling the methanol synthesis

Parameter	Unit	Value
Recycle ratio	mole/mole	3.5
Temperature at the reactor outlet	°C	255
Pressure	bar	60
Pressure drop in the reactor	bar	1
Stoichiometric number at the reactor inlet	mole/mole	2.07
Pressure level of the generated saturated steam	bar	40

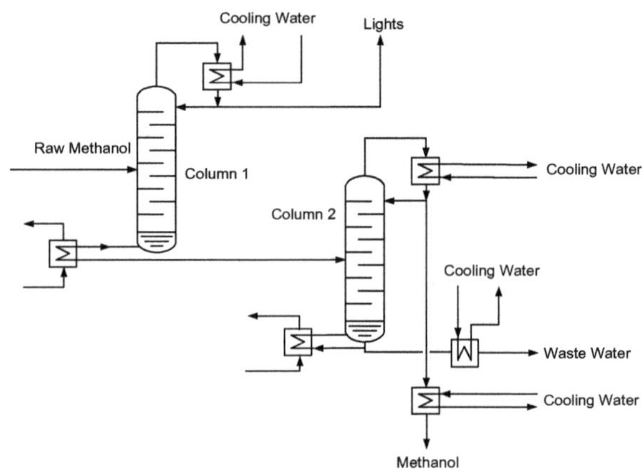

Fig. 6.7 Flow chart of methanol distillation

The product mixture after the methanol synthesis contains water and further impurities and requires a methanol distillation unit. Components with a lower boiling point, e.g. light gases (CO, CO_2, H_2, CH_4, N_2, etc.), dimethyl ether, ketones, formates, aldehydes, and with a higher boiling point, e.g. water, long-chain hydrocarbons, carbon acids, ethanol and higher alcohols, were separated from the methanol in a distillation column.

In commercial plants, the methanol purification step is performed using two or three distillation columns, depending on the boundary conditions. Both options lead to the same methanol purity. The light-boiling fraction is separated in the first distillation column, and the bottom product is fed to the second distillation column, where pure methanol is separated at the top of the column. The three-column design is used for methanol plant with a higher annual capacity. For better heat integration, the specific steam demand can be reduced by 60% in the case of the three-column option, but the CAPEX is higher. In general, methanol plants with a lower annual capacity, as considered in this study, use the two-column design. The flow chart of the modelled scheme is given in Fig. 6.7. Basic boundary conditions are listed in Table 6.10.

Table 6.10 Basic boundary conditions for modelling the methanol purification process

Parameter	Unit	Value
Pressure in the flash chamber	bar	5
Pressure in the first distillation column	bar	3
Stages in the first distillation column		20
Pressure in the second distillation column	bar	1.013
Stages in the second distillation column		30
Steam pressure for heating the distillation column	bar	5

6.4 Gasoline Synthesis

Several technologies were available for methanol-based gasoline synthesis, as described in Sect. 5.2.3. In this study, the single-stage MtG process by Mobil was modelled. In the first reaction step, methanol undergoes dehydration to form DME. Untreated methanol (raw methanol with 90 wt% methanol content) can be used as a feedstock. Consequently, the methanol purification step described in Sect. 6.3 can be omitted, making the overall process chain more feasible. Dehydration proceeds in an adiabatic fixed-bed reactor with Al_2O_3 as a catalyst. The uncooled reaction product is mixed with recycled unreacted DME (in order to fix the inlet temperature at 320–340 °C) and supplied to the gasoline converter. The MtG process uses adiabatic fixed-bed reactors with ZSM-5 type zeolite catalysts. At the reactor outlet, the reaction mixture is cooled in heat exchangers under steam generation and is separated into gaseous products, water and a hydrocarbon-rich fraction. After that, the hydrocarbon fraction is separated into different product fractions using four

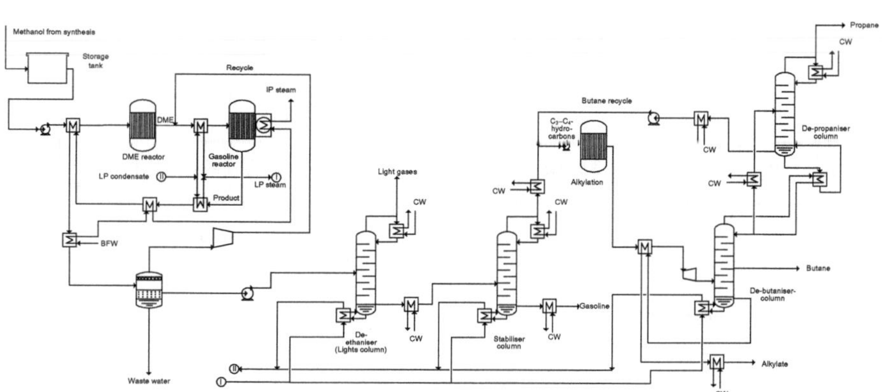

Fig. 6.8 Flow chart of MtG synthesis

Table 6.11 Basic boundary conditions for modelling MtG synthesis

Parameter	Unit	Value
Inlet temperature for the DME converter	°C	300
Pressure in the DME converter	bar	22
Inlet temperature for the gasoline converter	°C	330
Pressure in the gasoline converter	bar	21
Inlet temperature for the alkylation reactor	°C	40
Pressure in the alkylation reactor	bar	14

distillation columns. Light products (LPG) are separated at the top of the first column. The obtained gasoline fraction undergoes further treatment, including alkylation and isomerization steps. The flow chart of the MtG synthesis with the subsequent product treatment is shown in Fig. 6.8. Basic boundary conditions are listed in Table 6.11.

6.5 Overall Process Chains

The process chains described in the Sects. 6.1–6.4 are summarized below in flow charts. The steam production and steam integration between the particular processes is marked in blue. The flow chart for the overall process chain based on Siemens entrained-flow gasification is given in Fig. 6.9. The interaction between the individual processes in the COORVED-based process chain is illustrated in Fig. 6.10. To exemplify the benefits of the COORVED gasifier in comparison with a

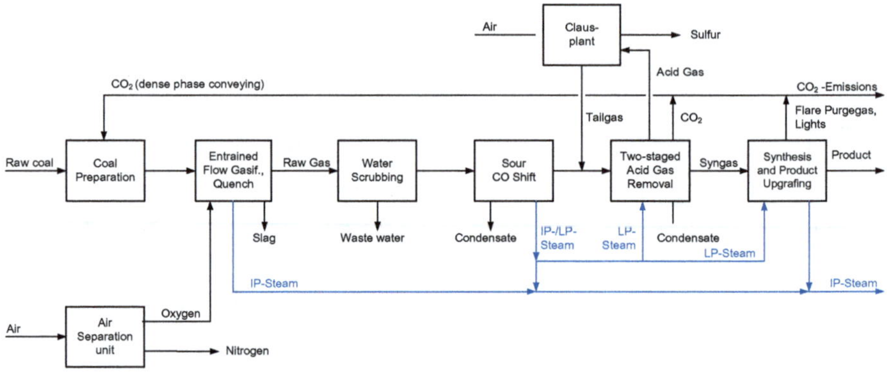

Fig. 6.9 Flow chart for the overall process chain based on the Siemens entrained-flow gasifier

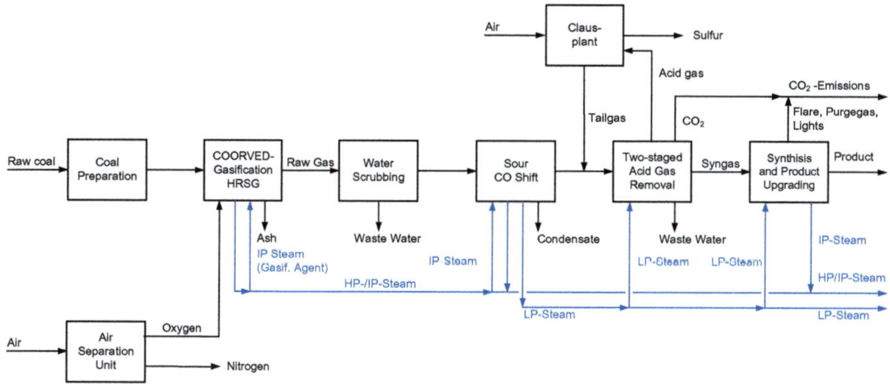

Fig. 6.10 Flow chart for the overall process chain based on the COORVED gasifier

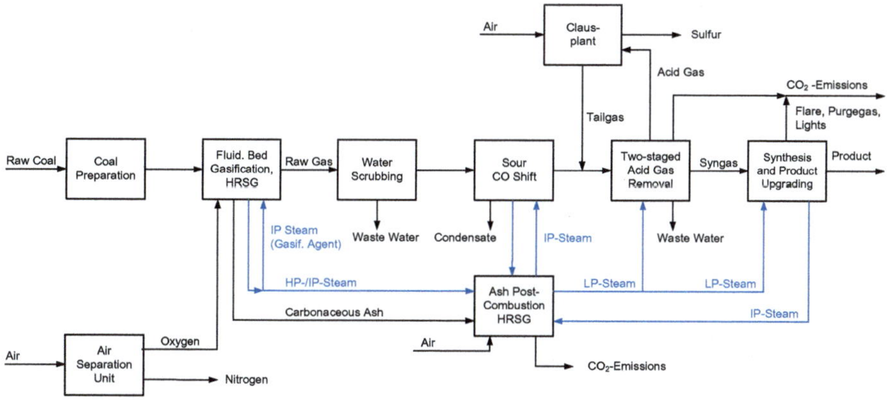

Fig. 6.11 Flow chart for the overall process chain based on the HTW fluidized-bed gasifier

conventional fluidized-bed gasifier (lower CAPEX, since some units can be omitted)—benefits which are discussed in Sect. 6.1—the flow chart for a process chain with fluidized-bed gasification is shown in Fig. 6.11.

Chapter 7
Technical and Economic Assessment

The overall process chains described in Chap. 6 are evaluated taking into account technological and economic aspects. The selection and nomenclature of the calculation case is represented in Fig. 6.1.

7.1 Technical Assessment

Basically, the comparison of the gasification processes itself is the starting point for the comparison of the overall process chains. The selection of process parameters is governed by the appropriate coal quality. The fundamental characteristics of both coals (TA and KO) and the gasification parameters selected for modelling (temperature, pressure, oxygen and steam demand) are summarized in Table 7.1. Furthermore, the calculated values of the specific assessment criteria (carbon conversion, syngas yield, etc.) and the raw gas composition are also listed.

Simulations verify the numerous advantages of the COORVED process in comparison with the conventional Siemens technology. The benefits are mainly due to the relatively low process temperature typical of the COORVED gasifier. Therefore, the COORVED process is characterized by a comparably low energy demand for the required sensible heat. The entrained-flow gasifiers operate at temperatures roughly 150 K above the ash flow temperature, in order to ensure that the ash fusion is complete. Despite the low gasification temperature, virtually complete carbon conversion can be achieved using COORVED gasification (typically 1.2 wt% C content in the bottom ash). In the hot zone, where fresh coal particles enter the gasifier and are contacted with the primary gasification agent, local temperatures are higher than in the rest of the gasifier. However, the ash flow temperature is not exceeded. Thus, the COORVED process has advantages over entrained-flow gasification, which is characterized by liquid slag formation. Due to these specifics, the oxygen demand for COORVED gasification is lower than for Siemens gasification by 18% for Taldinskaya coal and 22% for Komsomolets coal.

V. Litvinenko and B. Meyer, *Syngas Production: Status and Potential for Implementation in Russian Industry*,
https://doi.org/10.1007/978-3-319-70963-5_7

Table 7.1 Input data and calculation results for the gasification process

Coal		TA		KO	
Gasification technology		Siemens	COORVED	Siemens	COORVED
Parameter	Unit	Value			
Softening temperature	°C	1440	1440	1310	1310
Flow temperature	°C	1500	1500	1410	1410
Temperature at reactor outlet	°C	1650	1200	1550	1200
Pressure	bar	40	40	40	40
O_2 demand	m^3 (STP)/kg coal (waf)	0.711	0.583	0.774	0.606
Steam demand	kg/kg coal (waf)	<0.01	0.361	<0.01	0.379
Cold gas efficiency	%	77.7	84.9	75.8	84.2
Carbon conversion	%	99.5	99.8	99.5	99.2
Syngas yield	m^3 (STP) H_2 + CO/kg coal (waf)	2.02	2.21	2.13	2.37
H_2/CO	mole/mole	0.315	0.649	0.362	0.747
Raw gas composition	ex gasifier				
H_2O	vol.% (dry)	9.4	10.0	11.4	10.4
H_2	vol.% (dry)	21.8	35.9	23.8	39.3
CO	vol.% (dry)	69.3	55.4	65.8	52.6
CO_2	vol.% (dry)	7.5	7.2	8.9	6.6
CH_4	vol.% (dry)	0.0	0.2	0.0	0.3

This is of special importance for ash-rich coal (KO), because the COORVED process does not require the ash fusion to be complete. For example, it was noted that in the case of the ash melting enthalpy of 680 kJ/kg more than 1% of the overall combustion heat of the coal feed is required for the ash fusion to be complete. The COORVED process is also characterized by a higher syngas yield, thanks to a lower CO_2 production. This is caused by a lower demand for coal combustion heat for endothermic gasification reactions. Besides the better syngas yield, the COORVED process is also favoured compared to the Siemens entrained-flow technology because of the H_2/CO ratio in the syngas produced, which is twice as high. This is very beneficial, especially from the viewpoint of syngas utilization for methanol synthesis or liquid fuel production. The required size of the CO-shift reactor is therefore lower in the case of the process chains based on the COORVED process. At the same time, more of the coal carbon remains in the syngas as CO. Consequently, the yield of methanol or synthetic gasoline increases. The higher specific yield of the methanol or liquid fuel production effectuates a higher thermal efficiency, a better carbon integration rate and lower CO_2 emissions.

As expected, the cold gas efficiency for the ash-rich coal (KO) is lower than for the low-ash coal (TA). It is noticeable that this difference is especially high for the

KO-SIEMENS calculation case, despite the fact that the gasification temperature for this case is fixed 100 K lower than for the TA-SIEMENS case. Nevertheless, the low-quality KO coal is distinguished by a higher syngas yield. This applies both to the COORVED gasifier and the Siemens gasifier. These results could be explained by the higher H/C ratio and the lower oxygen content of the KO coal (see Table 2.1).

Three specific criteria, which are defined by Eqs. 7.1–7.3, were considered for the technical assessment of the overall process chains. For the estimation of the thermal efficiency (Eq. 7.2) it was supposed that the waste products (e.g. off-gases) could be utilized to produce energy by means of steam generation (efficiency factor of 90%). The thermal power of the coal feed (net calorific value of the coal mass flow) was converted into electric power to operate auxiliary units with an efficiency factor of 38.5%.

$$\text{Chemical efficiency: } \eta_{\text{stofflich}} = \frac{\dot{m}_{\text{MeOH}} \cdot Hu_{\text{MeOH}}}{\dot{m}_{\text{Kohle,Vergasung}} \cdot Hu_{\text{Kohle}}} \quad (7.1)$$

$$\text{Thermal efficiency: } \eta_{\text{th}} = \frac{\dot{m}_{\text{MeOH}} \cdot Hu_{\text{MeOH}} + 0,9 \cdot \dot{m}_{\text{Abprodukt}} \cdot Hu_{\text{Abprodukt}} + P_{\text{th,Überschussdampf}}}{\dot{m}_{\text{Kohle,Vergasung}} \cdot Hu_{\text{Kohle}} + \left(\frac{P_{\text{el,Hilfsenergie}}}{\eta_{\text{el,Netto,DKW}}}\right)} \quad (7.2)$$

$$\text{Carbon integration rate: } \eta_{\text{CR}} = \frac{\dot{n}_{\text{C,MeOH}}}{\dot{n}_{\text{C,Kohle,Vergasung}}} \quad (7.3)$$

The results of the calculated energy and material balances for the four overall process chains based on the consideration of the methanol synthesis are given in Table 7.2, and the results for the four process chains with a subsequent gasoline synthesis in Table 7.3. All eight calculation cases were modelled assuming a fuel-based thermal input power of 500 MW(th). It is evident that process chains which include COORVED gasification display distinct benefits compared to process chains including entrained-flow gasification. The expenses for the coal crushing are higher in the case of the Siemens process compared to the COORVED process. The calculations of the auxiliary power needs for the Siemens process take into account the CO_2 compression required for the dense coal feed (from atmospheric to >43 bar). In summary, the total energy demand for the Siemens process is 17–20% higher than for the COORVED process. The difference between the two gasification technologies is even higher, if specific values relating to the product output are considered: the specific energy demand for the Siemens process is 26–30% higher than for the COORVED process. This is caused by the higher methanol output in the case of process chains, which include COORVED gasification.

All process chains which include COORVED gasification are characterized by higher values for the chemical efficiency, thermal efficiency and carbon integration rate than the process chains with entrained-flow gasification. In the case of

Table 7.2 Energy and material balances for the process chains based on methanol synthesis

Coal		TA		KO	
Gasification technology		Siemens	COORVED	Siemens	COORVED
Parameter	Unit	Value			
Coal input	MW(th) kg/s t/d 10^3 t/a	500 20.9 1811 661.0	500 20.9 1811 661.0	500 25.4 2192 800.1	500 25.4 2192 800.1
Methanol output	MW(th) kg/s t/d 10^3 t/a	280.6 14.1 1217 444.7	314.2 15.8 1363 498.3	274.0 13.8 1189 435.2	312.8 15.7 1357 495.1
Product yield	kg MeOH/kg raw coal	0.675	0.756	0.543	0.618
Auxiliary power demand – Coal preparation – Air separation – Gasification – Gas cleaning – Methanol synthesis Overall	 MW(el) MW(el) MW(el) MW(el) MW(el) MW(el)	 1.13 20.69 4.43 4.93 3.63 34.81	 0.57 17.11 2.58 4.58 4.01 28.83	 1.37 20.83 4.43 4.76 3.70 35.09	 0.69 16.46 2.58 4.28 3.97 27.98
Specific energy demand	MJ/kg MeOH	2.47	1.83	2.55	1.78
Combustible waste products	MW(th)	23.78	25.76	21.70	24.19
Steam generation – High-pressure steam – Intermediate/low-pr. steam – Waste product utilization Overall	 MW(th) MW(th) MW(th) MW(th)	 0.00 112.30 21.40 133.70	 42.61 33.28 23.19 99.07	 0.00 118.43 19.53 137.96	 43.45 33.76 21.77 98.98
CO_2 emissions – Acid gas removal – Purge gas – Light gases from synthesis Overall	 kg/s kg/s kg/s kg/s	 25.08 1.13 1.41 27.62	 23.34 1.07 1.44 25.85	 22.73 1.02 1.37 25.11	 20.43 0.96 1.38 22.77
Specific CO_2 emissions	kg/kg MeOH	1.96	1.64	1.83	1.45
Chemical efficiency	%	56.1	62.8	54.8	62.6
Thermal efficiency	%	69.1	71.0	68.6	71.0
Carbon integration rate	%	40.4	45.3	42.0	48.0

Table 7.3 Energy and material balances for the process chains based on gasoline synthesis

Coal		TA		KO	
Gasification technology		Siemens	COORVED	Siemens	COORVED
Parameter	Unit	Value			
Coal input	MW(th) kg/s t/d 10^3 t/a	500 20.9 1811 661.0	500 20.9 1811 661.0	500 25.4 2192 800.1	500 25.4 2192 800.1
Gasoline output (0.31 kg gasoline per kg raw methanol)	MW(th) kg/s t/d 10^3 t/a	255.3 4.9 423 154.5	289.8 5.6 484 176.6	249.7 4.8 423 151.4	289.8 5.6 484 176.6
Product yield	kg gasoline/kg raw MeOH kg gasoline/kg raw coal	0.310 0.259	0.310 0.294	0.310 0.208	0.310 0.242
By-products – Alkylated – Propane – Butane	 kg/s kg/s kg/s	 0.17 0.30 0.51	 0.20 0.34 0.58	 0.17 0.29 0.50	 0.20 0.34 0.58
Auxiliary power demand – Coal preparation – Air separation – Gasification – Gas cleaning – Methanol synthesis – MtG synthesis Overall	 MW(el) MW(el) MW(el) MW(el) MW(el) MW(el) MW(el)	 1.13 20.69 4.43 4.93 3.63 1.22 36.03	 0.57 17.11 2.58 4.58 4.01 1.23 30.06	 1.37 20.83 4.43 4.76 3.70 1.22 36.31	 0.69 16.46 2.58 4.28 3.97 1.23 29.20
Specific energy demand	MJ/kg gasoline	6.13	4.50	6.31	4.37
Combustible waste products	MW(th)	23.78	25.76	21.70	24.19
Steam generation – High-pressure steam – Intermediate/ low-pr. steam – Waste product utilization Overall	 MW(th) MW(th) MW(th) MW(th)	 0.00 106.06 28.39 134.99	 42.61 67.92 28.37 138.9	 0.00 110.86 26.19 137.05	 43.45 68.72 26.31 138.48
CO_2 emissions – Acid gas removal – Purge gas – MtG synthesis Overall	 kg/s kg/s kg/s kg/s	 25.08 1.13 1.14 27.35	 23.34 1.07 1.29 25.70	 22.73 1.02 1.11 24.86	 20.43 0.96 1.29 22.68

(continued)

Table 7.3 (continued)

Coal		TA		KO	
Gasification technology		Siemens	COORVED	Siemens	COORVED
Parameter	Unit	Value			
Specific CO_2 emissions	kg/kg products	4.65	3.85	4.32	3.40
Chemical efficiency	%	51.0	58.0	49.9	58.0
Thermal efficiency	%	64.3	65.4	63.6	65.5
Carbon integration rate	%	38.4	43.6	40.0	46.4

COORVED gasification, steam production is initially higher than for Siemens gasification, due to the use of convective heat exchangers instead of water quenching. However, some of the steam produced is required for the sour gas shift process in the COORVED-based process chains, in contrast to the Siemens-based process chains. Altogether, the total steam output in the case of COORVED-based process chains is lower than for the Siemens-based process chains. Nevertheless, the process chains which include COORVED gasification are characterized by slightly higher thermal efficiency values compared to the process chains which include Siemens gasification.

7.2 Economic Assessment

Using the modelling data for the overall process chains (calculated energy and material balances), an economic assessment of the process chains was performed. The estimation of CAPEX costs is based on the capacity rating method. Product costs were calculated as an average over the whole payback period. Major boundary conditions and assumptions are summarized in Tables 7.4 and 7.5.

Information about the reference plants, including CAPEX costs and specifications of the material flows used for scaling, is given in Table 7.6.

The assessment of CAPEX for the COORVED gasifier is based on the corresponding data for the commercial fluidized-bed HTW gasifier. Thanks to the smaller reactor size, the CAPEX for the COORVED gasifier can be reduced by 30% compared to the HTW gasifier. On the other hand, COORVED gasification has so far only been demonstrated on a pilot scale. For this reason, a technology risk factor of 1.2 was taken into account for the COORVED gasifier, as is usual in such cases.

The resulting CAPEX costs are summarized for the process chains based on methanol synthesis in Table 7.7 and for the process chains with subsequent gasoline production in Table 7.8. The contributions of the main plant components to the overall CAPEX are also listed.

Table 7.4 Boundary conditions for the economic assessment [79, 80]

Parameter	Value
General boundary conditions	
Reference year	2015
Start of the project planning	01.01.2016
Start of the plant construction	01.01.2017
Start of the plant operation	01.01.2020
Payback period	20 a
Rate of inflation	2% p.a.
Price rise	1% p.a.
Financing structure and interest rate	
Shareholder equity	33%
Interest rate of the shareholders	17.5%
Credit capital	67%
Interest rate of the credit capital	6.7%
Resulting internal rate of interest	10% p.a.
Required CAPEX during the plant construction	
2017–2018	10%
2018–2019	60%
2019–2020	30%
Fixed OPEX (relating to the overall CAPEX)	
Insurance	0.5%
Service and maintenance	5%
Personnel costs per employee	86,208 € p.a.

Table 7.5 Assumptions concerning the costs and revenues [79–81]

Parameter	Reference year	Unit	Value
Variable costs			
Komsomolets coal	2015	€/t	20.00
Taldinskaya coal	2015	€/t	25.00
Electricity	2015	€/MWh	42.00
Feed water	2008	€/t	0.20
Costs for disposal of waste			
Waste water	2015	€/t	0.74
Ash/Slag	2008	€/t	0.00
CO_2 penalties	2015	€/t	10.00
Revenues for by-products			
Sulphur	2010	€/t	149.60
Propane/Butane	2015	€/t	553.40

Table 7.6 Reference data for the estimation of CAPEX costs according to the capacity rating method [79, 80, 82]

Main plant components	Material flow used for scaling	Costs (million €)	Capacity (kg/s)
Siemens gasifier	Input raw coal	192	40.0
HTW gasifier	Input raw coal	248	46.0
COORVED gasifier	Input raw coal	208	46.0
Air separation unit	Oxygen	180	95.4
Acid gas removal unit	Separated CO_2	101	181.9
Methanol synthesis	Syngas feed	6	35.6
Methanol purification	Raw methanol	3	11.2
MtG synthesis	Raw methanol	4	10.6

Table 7.7 Overall CAPEX in million € for the process chains based on the methanol synthesis

Calculation case	TA-SIEMENS MeOH	TA-COORVED MeOH	KO-SIEMENS MeOH	KO-COORVED MeOH
Coal preparation	68.114	65.264	77.843	74.587
Air separation	59.848	52.383	60.138	51.000
Gasifier island	128.969	126.381	147.390	144.435
Gas cleaning with piping	62.150	51.318	63.453	50.299
Synthesis and product treatment	9.183	9.705	9.040	9.638
Infrastructure with attachments	44.018	22.482	50.305	25.693
Onsite costs	372.282	327.532	408.170	355.651
Overall CAPEX	542.871	478.724	594.477	519.627

The utilization of the ash-rich KO coal requires a higher CAPEX compared to the low-ash TA coal. As regards the comparison between the gasification technologies, the process chains based on the Siemens gasification require 5–7% higher CAPEX than the equivalent process chains with COORVED gasification. The contributions of the plant components to the overall CAPEX can be recognized better if technological aspects are taken into consideration. The more expensive coal preparation in the case of the Siemens gasification leads to a 4% greater CAPEX. The contribution of the air separation unit is 12–15% higher for Siemens gasification due to the higher oxygen demand. Despite the technology risk factor, the CAPEX for the COORVED gasifier is lower than for the Siemens gasifier. In addition, further economic benefits arise from the easier ash discharge system (taken into account in the line "Infrastructure with attachments") and less expensive

Table 7.8 Overall CAPEX in million € for the process chains based on gasoline production

Calculation case	TA-SIEMENS MtG	TA-COORVED MtG	KO-SIEMENS MtG	KO-COORVED MtG
Coal preparation	68.114	65.264	77.843	74.587
Air separation	59.848	52.383	60.138	51.000
Gasifier island	128.969	126.381	147.390	144.435
Gas cleaning with piping	62.150	51.318	63.452	50.299
Synthesis and product treatment	35.409	38.000	34.894	37.958
Infrastructure with attachments	44.018	22.482	50.305	25.693
Onsite costs	398.507	355.827	434.024	383.971
Overall CAPEX	580.321	518.733	631.347	558.555

gas cleaning process in the case of COORVED gasification. Only the CAPEX for synthesis and product treatment is higher for process chains with COORVED gasification. But this fact is caused by the higher syngas output and therefore higher product yield (methanol, gasoline) in process chains based on the COORVED gasification.

Estimated CAPEX costs are used to calculate the averaged product costs, and the results obtained are listed in Tables 7.9 (methanol production) and 7.10 (gasoline production). The contribution of various types of costs is also represented, including the revenues for the by-products.

As expected, product costs are lower for the utilization of the low-ash TA coal than for the high-ash KO coal, similar to equivalent trends for CAPEX. For the calculations, it was assumed that the price for TA coal is 5 €/t higher than for KO coal. Nevertheless, the contribution of the fuel cost to the final product cost is lower in the case of the TA coal, due to the higher product yield. In addition, TA coal benefits from the lower OPEX costs, which are enabled thanks to the reduced effort for syngas cleaning. A comparison of the gasification technologies reveals a lower product cost for process chains based on COORVED gasification. The driving forces are the lower CAPEX, higher product yield (and consequently lower contribution of the fuel costs) and reduced effort for gas cleaning (lower OPEX). On the other hand, the revenues from the by-products are higher in the case of Siemens gasification, due to the higher steam export. But this contribution does not outweigh other costs.

It can be further noted that process chains with gasoline production stand out for having significantly higher revenues from by-products. However, the additional methanol conversion process causes additional costs.

Table 7.9 Averaged product cost in € for the process chains based on methanol synthesis

Calculation case	TA-SIEMENS MeOH	TA-COORVED MeOH	KO-SIEMENS MeOH	KO-COORVED MeOH
CAPEX	34.30	27.01	38.47	29.44
Fuel costs	7.70	6.88	7.64	6.69
OPEX including service	19.99	16.47	22.27	18.26
Annual costs	62.00	50.36	68.38	54.39
Revenue for the by-products	−5.77	−3.97	−6.07	−3.99
Product costs				
– in €/MWh of product	56.24	46.39	62.31	50.40
– in €/t of product	310.86	256.45	344.44	278.62
– in €/litre of product	0.25	0.20	0.27	0.22

Table 7.10 Averaged product cost in € for the process chains based on gasoline production

Calculation case	TA-SIEMENS MtG	TA-COORVED MtG	KO-SIEMENS MtG	KO-COORVED MtG
CAPEX	40.31	31.73	44.84	34.17
Fuel costs	8.47	7.46	8.38	7.22
OPEX including service	22.66	17.71	25.01	17.79
Annual costs	71.44	56.90	78.23	59.18
Revenue for the by-products	−13.15	−10.91	−13.38	−10.86
Product costs				
– in €/MWh of product	58.29	45.99	64.85	48.32
– in €/t of product	697.17	550.01	775.63	577.93
– in €/litre of product	0.54	0.43	0.60	0.45

Table 7.11 Results of the sensitivity analysis for the calculation case TA-COORVED-MtG

Parameter	Unit	Lower value	Basic value	Upper value
CAPEX	millions of €	189.9 (−50%)	379.9	569.9 (+50%)
Averaged product costs	€/MWh	29.44	45.99	62.54
Plant availability	%	80 (−10%)	88.7	95 (+7%)
Averaged product costs	€/MWh	50.10	45.99	43.48
Revenue for LPG	€/t	276.7 (−50%)	553.4	830.1 (+50%)
Averaged product costs	€/MWh	49.37	45.99	42.60
Electricity	€/MWh	15 (−64%)	42	100 (+138%)
Averaged product costs	€/MWh	45.47	45.99	47.09

In order to identify the factors having the biggest possible effect on the product costs, a sensitivity analysis was performed in the present study. As an example, the TA-COORVED-MtG calculation case was considered and some influencing factors were varied in an appropriate range. Selected results were represented in Table 7.11. It is evident that the major influence of the product costs is due to CAPEX costs. The plant availability also strongly influences the product costs. Benefits from increased by-product revenues are recognisable only at relatively high by-product prices. Electricity costs have a minor influence.

Chapter 8
Summary

The application potential of different gasification technologies for the non-fuel utilization of Russian coals of varied quality (low-ash coals, ash-rich coals) was evaluated in this study. Based on this analysis, the two most suitable gasification processes were selected: the entrained-flow gasification process commercially offered by Siemens and the innovative COORVED process, which was developed especially for the utilization of low-quality coals. In addition, an overview of the technologically proven and economically most feasible routes for the syngas conversion was given and two most suitable products were selected: methanol and synthetic gasoline. The resulting overall process chains were modelled using the software package Aspen Plus® for two selected coals from the Kuzbass coal region—the low-ash Taldinskaya coal and the ash-rich Komsomolets coal. The estimated technical and economic criteria allow a well-grounded assessment of the application potentials of the non-fuel utilization of Russian coals taking into account the local boundary conditions of the coal regions (e.g. Kuzbass) and taking into account the basic characteristics of the investigated coals.

The calculation of the material balances leads to the following results:

- For the low-ash Taldinskaya coal

$$\begin{aligned} \text{Siemens process: 1 tonne raw coal} &\rightarrow 675\,\text{kg methanol} \rightarrow 259\,\text{kg gasoline} \\ \text{COORVED process: 1 tonne raw coal} &\rightarrow 756\,\text{kg methanol} \rightarrow 294\,\text{kg gasoline} \end{aligned}$$

- For the ash-rich Komsomolets coal

$$\begin{aligned} \text{Siemens process: 1 tonne raw coal} &\rightarrow 543\,\text{kg methanol} \rightarrow 208\,\text{kg gasoline} \\ \text{COORVED process: 1 tonne raw coal} &\rightarrow 618\,\text{kg methanol} \rightarrow 242\,\text{kg gasoline} \end{aligned}$$

V. Litvinenko and B. Meyer, *Syngas Production: Status and Potential for Implementation in Russian Industry*,
https://doi.org/10.1007/978-3-319-70963-5_8

In both cases (low-ash coal, ash-rich coal), the COORVED process has clear advantages compared to the commercial entrained-flow Siemens process.

Similar conclusions can be derived from economic consideration (product costs):

- For the low-ash Taldinskaya coal

Siemens process:	311 €/t methanol	0.54 €/l gasoline
COORVED process:	256 €/t methanol	0.43 €/l gasoline

- For the ash-rich Komsomolets coal

Siemens process:	344 €/t methanol	0.60 €/l gasoline
COORVED process:	279 €/t methanol	0.45 €/l gasoline

Calculations show that coal-based methanol production could be economically feasible under the boundary conditions of the Russian coal industry. An assessment of process chains based on the utilization of the innovative COORVED gasification reveals product costs for methanol of around 250 €/t for the low-ash coal and approximately 280 €/t for the ash-rich coal. The world methanol price varied in 2010–2015 in the range of 300–500 €/t. Only in 2016 did it fall below 250 €/t due to the decelerated economic development. Currently, the methanol price is rising again: in January 2017 it was approx. 370 €/t in Europe and approx. 430 €/t in Asia. In view of the geographical location of the major coal regions in Russia and possible access to the markets in Asia, it is interesting to note that methanol prices in Asia have already been higher than the methanol prices in Europe for the last 15 years (the difference was temporarily greater than 100 €/t).

However, it should be borne in mind that the calculations performed are based on some optimistic assumptions. For example, it was not taken into consideration that industrial-scale plants usually include redundant stand-by gasifiers in the budget, at the expense of higher CAPEX. In practice, an additional increase in CAPEX by roughly 10–20% can be expected. From this point of view, the Annex concepts which were described in this study are of particular interest. They can potentially reduce costs by 10–15%.

In conclusion, this study demonstrates that there is considerable potential for the non-fuel utilization of Russian coals, including low-quality ash-rich coals, provided that advanced, modern technologies are used for coal gasification. Nevertheless, the preliminary assessment performed in this study points out that production costs for coal-based products are near the limit of profitability. In this context, it seems reasonable to envisage a further, more detailed economic evaluation in future. It is preferable to implement such an analysis in application to a specially selected site in order to account precisely for the local boundary conditions and the potential key markets. In addition, the potential of feasible Annex concepts can be quantified.

Appendix

Производство синтез-газа: современное состояние и возможности внедрения в российской промышленности

В.С. Литвиненко, Б. Майер

Содержание

V. Litvinenko and B. Meyer, *Syngas Production: Status and Potential for Implementation in Russian Industry*,
https://doi.org/10.1007/978-3-319-70963-5

Краткое резюме

В данном научно-техническом исследовании проведен анализ применимости различных технологий газификации к глубокой переработке российских углей. Особое внимание уделено выбору экономически целесообразных схем получения востребованных на рынке продуктов из высокозольных углей, имеющих ограниченный потенциал реализации в качестве энергетического угля. На сегодняшний день наиболее приемлимыми для переработки высокозольных углей являются газогенераторы с подвижным слоем. Они работают при относительно низких температурах и отличаются пониженным удельным потреблением кислорода. Однако состав полученного синтез-газа непригоден для переработки в востребованные продукты (например в метанол, этилен, аммиак, синтетическое жидкое топливо). Газогенераторы с кипящим слоем также работают при температуре ниже точки плавления золы, благодаря чему отпадает необходимость в подводе тепла для расплавления золы. Принципиальным недостатком данной технологии является недостаточно высокая степень конверсии углерода (выведенная из газогенератора зола содержит до 25 вес.-% остаточного углерода). Большинство коммерческих газогенераторов использует технологию поточной газификации. Она обеспечивает высокую степень конверсии углерода и особо высокое качество синтез-газа, наилучшим образом подходящее для конверсии в конечный продукт. Но поточные газогенераторы неприменимы для экономически рентабельной глубокой переработки углей с высоким содержанием золы, так как они работают при температурах выше 1.200 °C и требуют в случае высокозольных углей подвода значительного количества тепла, необходимого для нагрева и расплавления золы. В данном исследовании проведено моделирование поточной технологии СИМЕНС с сухим подводом угля. Эта технология имеет преимущества по сравнению с конкурирующими технологиями, прежде всего благодаря значительно более низким капитальным вложениям.

Учитывая недостатки применения коммерческих технологий к переработке высокозольных углей, дополнительно рассмотрена новая технология COORVED, оптимально приспособленная именно к подобным, «сложным» углям. В рамках этой технологии реализована концепция многозонного газогенератора, комбинирующего преимущества газогенераторов с подвижным слоем (оптимальная утилизация тепла), с кипящим слоем (невысокая удельная потребность в кислороде) и поточных газогенераторов (высокое качество синтез-газа).

При помощи коммерческого программного пакета Aspen Plus® проведен анализ полных технологических цепочек глубокой переработки для двух кузбасских углей – низкозольного угля (шахта «Талдинская-Западная-2») и высокозольного угля (шахта «Комсомолец»). Показано, что наиболее пригодной является технология COORVED, имеющая преимущества по сравнению с коммерческими технологиями как с точки зрения энергетического и материального баланса, так и в отношении экономических показателей. В конечном итоге, расчеты цепочек привели к следующим значениям себестоимости продукта, полученного технологией COORVED:

шахта «Талдинская-Западная-2»:
256 евро/тонну метанола 0,43 евро/литр бензина

шахта «Комсомолец»:
279 евро/тонну метанола 0,45 евро/литр бензина

Данные показывают, что производство метанола из российских углей на базе передовых технологий газификации может быть вполне конкурентоспособным. Цена метанола на мировых рынках составляла в последние годы 300–500 евро/тонну и лишь в 2016 году опустилась, в связи с экономическим кризисом, ниже 250 евро/тонну. На начало 2017 года цена на метанол опять превышает 300 евро/тонну. В отличие от метанола, себестоимость производства синтетического жидкого топлива из угля лежит над уровнем рыночных цен. В этой связи, глубокая переработка угля с получением СЖТ представляется целесообразной лишь в труднодоступных регионах с богатыми залежами угля, например на Дальнем Востоке.

1. Введение

По сравнению с другими горючими полезными ископаемыми, уголь отличается более равномерным распределением месторождений и запасов по всему миру. Однако при этом наблюдаются значительные отличия относительно качества угля. Возможности использования угля для электрогенерации и производства химических продуктов в существенной мере определяются характеристиками перерабатываемого сырья. Базовым процессом для цепочек по переработке угля в товарную продукцию является газификация угля с получением синтез-газа. Выбор подходящих технологий для реализации намеченных цепочек переработки угля в значительной степени зависит от рамочных условий для угледобывающего и углеперерабатывающего производства и от потенциальных рынков сбыта конечной продукции.

В рамках данного научно-технического исследования фокус положен на оценку возможностей экономически целесообразной переработки «сложных» углей, с высоким содержанием золы и низкой удельной теплотворной способностью. Эффективным вариантом использования подобных углей представляется переработка в конечный продукт с высокой добавленной стоимостью. В условиях российской промышленности дополнительным аспектом является географическое положение угольных регионов, расположенных далеко от промышленных центров России, так что транспортные расходы налагают определенные ограничения на рентабельность сбыта товарного угля.

Проведенное научно-техническое исследование включает в себя несколько этапов, представленных в данном отчете. Вначале будут рассмотрены различные технологии газификации угля и обсуждены требования к углю, определяющие выбор наиболее подходящей технологии переработки. Дополнительно будут коротко представлены концепции, основанные на использовании эффектов синергии при комбинации углехимического производства с расположенной в непосредственной близости угольной электростанцией (углехимический-энергетический кластер). Подобные подходы позволяют существенно сократить объем необходимых капитальных вложений и эксплуатационных расходов. Далее будут представлены экономически наиболее значимые технологии переработки полученного синтез-газа в конечную продукцию. Рассмотренные технологии получения и переработки синтез-газа проанализированы, с учетом технических и экономических показателей, с точки зрения их применимости к российским углям. На основе этого анализа были выбраны наиболее приемлимые технологические цепочки, для которых проведены расчеты массового и энергетического баланса, а также экономических показателей. В заключении, на базе рассчитанных показателей дана техническая и экономическая оценка потенциала реализации рассмотренных цепочек от угля до конечного продукта. При этом особый упор сделан на глубокую переработку высокозольных углей, для которых, ввиду высокой доли транспортных расходов в структуре ценообразования, наиболее перспективным подходом представляется переработка на месте, с получением продукции, востребованной на российском и мировом рынке.

2. **Выбор углей**

Исходным пунктом для данного исследования явились данные по техническим характеристикам российских углей из десяти угольных шахт, расположенных в четырех угольных бассейнах [1]. Из них были выбраны два примера, которые оптимальным образом представляют весь спектр рассмотренных углей. Это низкозольный уголь шахты «Талдинская-Западная-2» и высокозольный уголь шахты «Комсомолец». Показатели, играющие определяющую роль при выборе оптимальной технологии переработки, собраны в Таблице 1.

Таблица 1 Показатели для углей, выбранных для оценки применимости для глубокой переработки [1]

Показатель	Ед. изм.	Талдинская	Комсомолец
Общий анализ			
Связанный углерод	вес.-% (сухой)	55,43	33,81
Выход летучих веществ	вес.-% (сухой)	30,97	30,66
Зольность	вес.-% (сухой)	13,61	35,53
Влага общая	вес.-%	11,80	9,65
Химический анализ			
C	вес.-% (сухой)	70,60	53,47
H	вес.-% (сухой)	3,97	3,82
O	вес.-% (сухой)	9,31	4,96
N	вес.-% (сухой)	2,15	1,59
S	вес.-% (сухой)	0,22	0,55
Cl + P + As	вес.-% (сухой)	0,15	0,10
Теплотворная способность (рядовой уголь)			
Низшая теплота сгорания	МДж/кг	23,85	19,71
Высшая теплота сгорания	МДж/кг	24,90	20,69
Плавкость золы (востанавливающая атмосфера)			
Температура начала деформации	°C	1440	1310
Температура размягчения	°C	1470	1360
Температура жидкого состояния	°C	1500	1410
Другие показатели			
Индекс свободного вспучивания	–	1	–
Размолоспособность по методу Хардгрова	–	58	57
Размер куска - более 50 мм - менее 2 мм	мм % %	0-200 – –	0-200 11,1 20,4
Химический состав золы			
SiO_2 Al_2O_3 Fe_2O_3 CaO MgO TiO_2 MnO_2 P_2O_5 SO_3 Na_2O K_2O	вес.-% вес.-% вес.-% вес.-% вес.-% вес.-% вес.-% вес.-% вес.-% вес.-% вес.-%	60,90 23,40 5,00 2,90 2,00 0,90 0,03 0,40 1,30 0,8 2,40	42,66 16,75 10,13 10,58 3,86 0,89 0,03 0,20 9,06 2,18 2,63

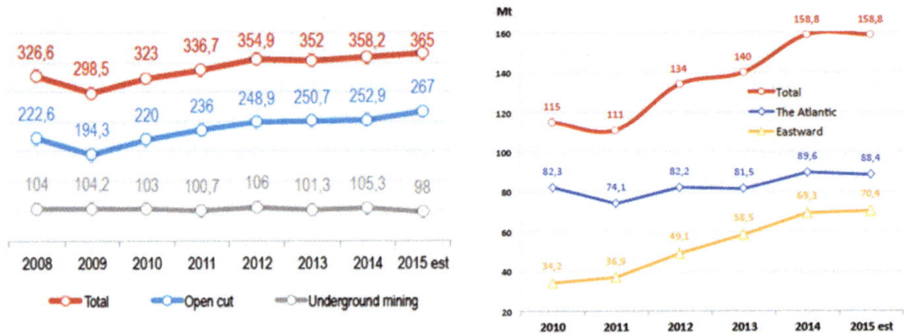

Рисунок 1 Объемы добычи угля (слева) и экспорта угля (справа) в России [4]

Россия располагает вторыми по объему запасами угля в мире (157 миллиардов тонн) бассейнах [2]. Угольная промышленность представляет собой одну из наиболее мощных отраслей экономики, которая была приватизирована в конце 90-ых годов и с тех пор подверглась коренной модернизации. Она является прибыльной отраслью, полностью обходящейся без государственных дотаций. Объемы годовой добычи угля неуклонно растут, см. Рисунок 1 (слева), и приближаются к уровню добычи в советские времена (1990: 395 миллионов тонн в год) [3]. Доля угля, идущего на экспорт, превышает 40% и также растет, как видно на Рисунке 1 (справа).

Ориентация на экспорт еще более ярко выражена в Кузбассе – самом крупном угольном бассейне России. Как видно на Рисунке 2, доля экспорта в

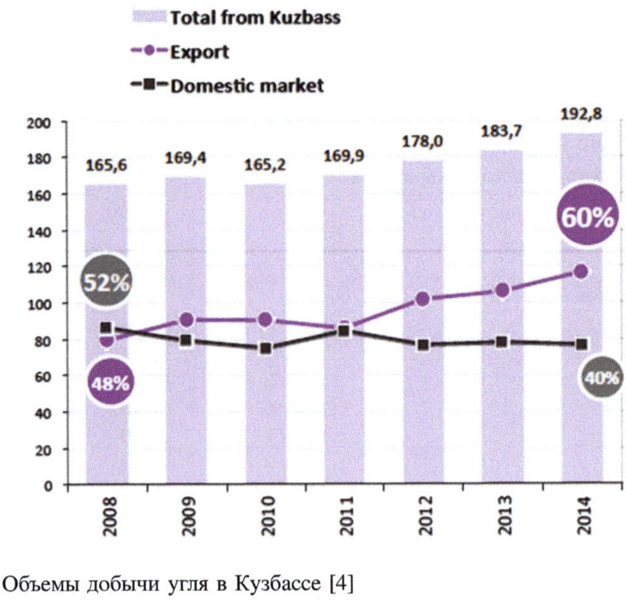

Рисунок 2 Объемы добычи угля в Кузбассе [4]

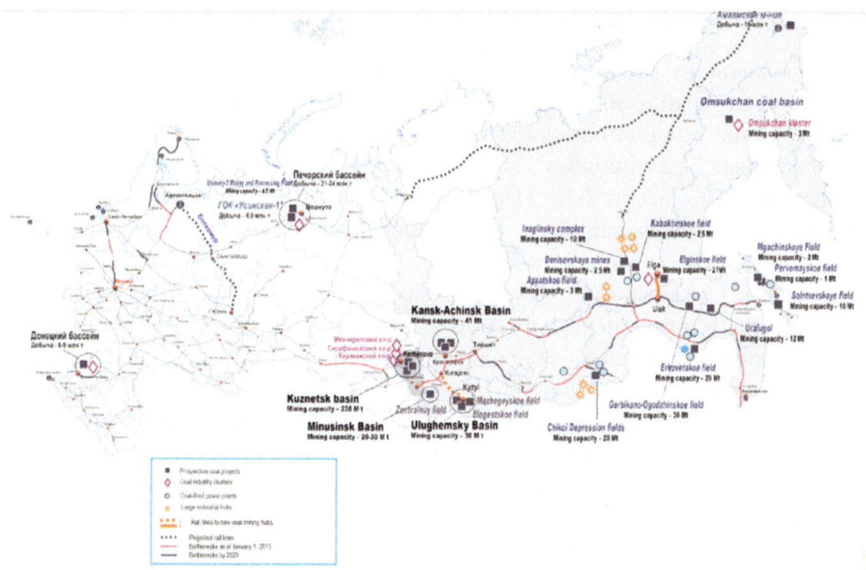

Рисунок 3 Добыча и транспорт угля в России [4]

общем объеме сбыта кузбасских углей выросла с 48% в 2008 году до 60% в 2014 году.

Транспортные пути из Кузбасса как до портов отгрузки на Дальнем Востоке, так и до потенциальных рынков сбыта в Европе и европейской части Российской Федерации составляют несколько тысяч километров. Другие угольные бассейны России также расположены в удаленных и зачастую труднодоступных регионах, см. Рисунок 3.

Значительная часть угольных месторождений в России приходится на высокозольные угли. Для таких «низкокачественных» углей глубокая переработка может представлять экономически более целесообразный вариант, чем их сбыт в качестве энергетического угля на мировых рынках. Ключевую роль для глубокой переработки углей играет процесс газификации с получением синтез-газа. В настоящее время в российской промышленности ощущается потребность и растущий интерес к внедрению передовых, экономически целесообразных технологий получения востребованных на рынке конечных продуктов. Реализация процессов глубокой переработки угля позволит укрепить экономическую независимость России от скачкообразного развития мировых цен на горючие полезные ископаемые и внести вклад в создание полных производственных цепочек с приростом добавочной стоимости внутри страны. Внедрение перспективных технологий глубокой переработки угля также будет способствовать долгосрочному сохранению рабочих мест в угольной промышленности и созданию новых, высококвалифицированных рабочих мест в углехимических кластерах.

3. Технологии газификации

Для газификации угля, очистки полученного синтез-газа и последующих процессов переработки в различные продукты (например метанол, аммиак, синтетическое топливо) имеется большое число коммерческих технологий, разработанных различными компаниями и успешно реализованных в промышленном масштабе. Первая задача – это подбор подходящей технологии газификации, наилучшим образом соответствующей характеристикам перерабатываемого угля. Различные газогенераторы существенно отличаются друг от друга относительно эффективности газификации в применении к выбранному виду угля. С другой стороны, газогенератор, оптимально пригодный для переработки определенного угля, не обязательно также подходит и для газификации угля с другими характеристиками. В этом разделе будут коротко представлены основные типы газогенераторов, при этом особое внимание будет уделено оценке их применимости к газификации углей шахты «Талдинская-Западная-2» и «Комсомолец». Уголь шахты «Комсомолец» с содержанием золы 35,5 вес.-% (на сухое состояние топлива) относится к высокозольным углям.

3.1. Сжигание и газификация

В обоих случаях, сжигания и газификации, кусок угля проходит одинаковые стадии конверсии: нагрев, обезвоживание, выход летучих веществ, реакция полученного кокса. После окончания процесса конверсии остается зола) [5]. Наиболее важным различием между сжиганием и газификацией является количество используемого кислорода и соответственно состав полученного газа. В процессе сжигания подводится избыток кислорода (обычно в виде воздуха), необходимый для полного окисления углеводородных составляющих угля. В результате процесса выделяется тепло и образуется азот N_2 (если в качестве окислителя использовался воздух), углекислый газ CO_2 и водяной пар H_2O. В угольных электростанциях тепло используется для электрогенерации, газ остается без применения. В процессе газификации подводится значительно меньшее количество кислорода (30–50% по отношению к сжиганию). Как правило, вместо воздуха используется чистый кислород. Часть угля подвергается парциальному окислению, с выделением тепла. Полученный газ состоит в основном из моноокиси углерода CO и водорода H_2. Около 70–85% теплотворной способности исходного угля сохранилось в полученном синтез-газе. Газ может быть подвергнут сжиганию в газовой турбине для генерации электроэнергии (парогазовая установка с внутрицикловой газификацией), но в большинстве случаев этот газ используется для синтеза химических продуктов или топлива. Таким образом газификация угля позволяет получить продукты, которые можно легче транспортировать и использовать, чем сам уголь [6].

3.2. Мировой рынок технологий газификации

На сегодняшний день в мире подвергается газификации около 170 миллионов тонн угля в год [7]. Это соответствует примерно половине годовой добычи угля в России. По прогнозам, ожидается удвоение мировых мощностей от 147 ГВт термической мощности в 2015 году до примерно 359 ГВт в 2020 году [7]. В настоящее время 65% произведенного синтез-газа перерабатывается в химические продукты (метанол, олефины, оксо-соединения, аммиак, мочевина, водород и др.), 18% в жидкое топливо (дизель, керосин, бензин и др.), 10% в газообразное топливо (синтетический природный газ) и 7% идет на электрогенерацию (ПГУ с внутрицикловой газификацией). Как показано на Рисунке 4, для газификации могут использоваться самые различные виды углеводородного сырья и могут быть получены разнообразные продукты. Однако уголь играет доминирующую роль среди видов сырья для газификации, его доля составляет на сегодняшний день 75,5%. В будущем ожидается дальнейший прирост, до 83% в 2020 году.

Предложен целый ряд коммерческих технологий газификации угля, некоторые из них представлены на Рисунке 5. Наиболее распространенные технологии собраны в Таблице 2, с указанием общих мощностей установок, находящихся в эксплуатации.

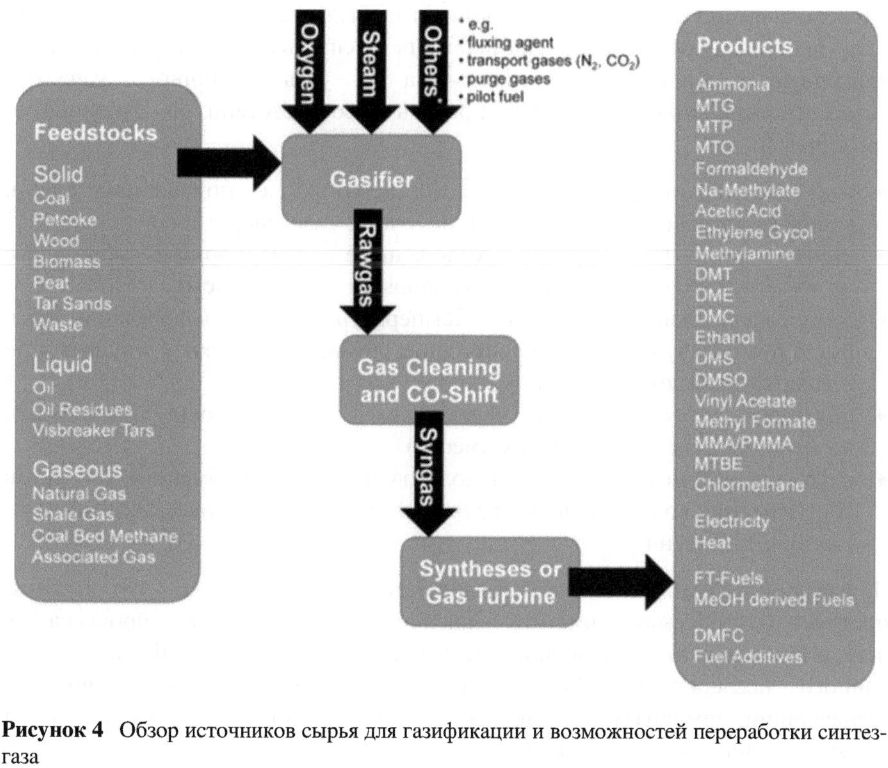

Рисунок 4 Обзор источников сырья для газификации и возможностей переработки синтез-газа

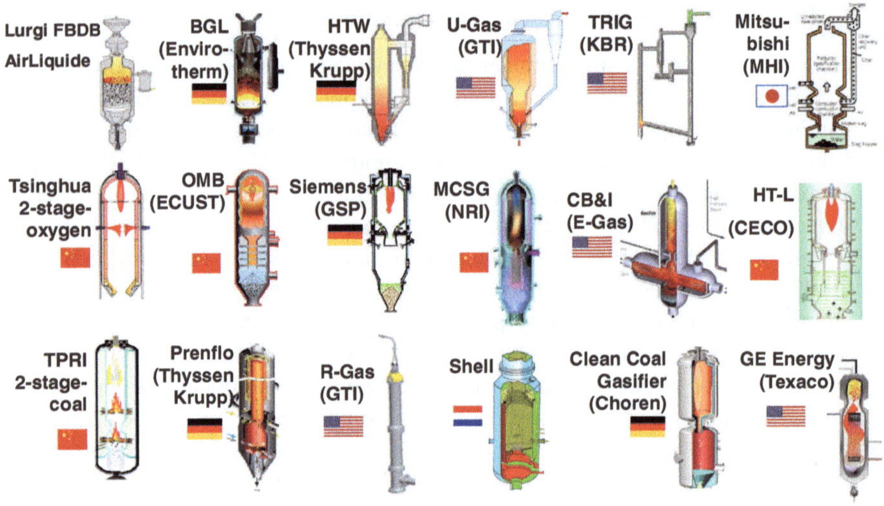

Рисунок 5 Обзор коммерческих технологий газификации

3.3. **Газогенераторы**

Технологические основы газификации угля были заложены в начале 20-го века. За прошедшие годы были разработаны различные типы газогенераторов, пригодные к производству синтез-газа из угля различного качества. Газогенераторы можно классифицировать в соответствии со следующими критериями:

- тип технологии газификации (газогенераторы с подвижным слоем, газогенераторы с кипящим слоем, поточные газогенераторы)
- стенки реактора (стенки выложенные огнеупорным кирпичем, мембранные стенки, стенки с водо-паровым охлаждением)
- температура газификации (ниже температуры начала деформации золы, выше температуры жидкого состояния золы, в некоторых случаях при температуре агломерации золы)
- подвод угля (сухой подвод через шлюзы, сухой подвод при помощи насосов, подвод водоугольных смесей)
- охлаждение синтез-газа (с использованием теплообменников, путем вбрызгивания воды – водной квенч, путем подвода химических веществ – химический квенч)

Другие показатели, по которым различаются газогенераторы – это давление процесса, требуемая чистота кислорода, проведение процесса с использованием катализаторов или газификация угля с добавкой флюсов. Для данного исследования определяющую роль играют тип технологии газификации, температура газификации и состав синтез-газа, на них будет сфокусировано особое внимание.

Таблица 2 Наиболее распространенные коммерческие технологии газификации [7]

Общая мощность в ГВт (терм.)	Эксплуат.	Строит.	Проект.	Доля в %	Доля в %
	2014	2016	2019	2014	2019
Дженерел Электрикс (поточная газификация, водоугольн. смеси)	30	43	46	25	19
Шелл (поточная газификация, сухой подвод угля)	30	32	39	25	16
Лурги FBDB (подвижный слой, сухой подвод угля)	19	20	21	16	9
OMB (поточная газификация, водоугольные смеси)	7	18	20	6	8
SEDIN (подвижный слой, сухой подвод угля)	6	11	20	5	8
Сименс (поточная газификация, сухой подвод угля)	2	14	17	2	7
E-Gas (поточная газификация, водоугольные смеси)	1	11	17	1	7
HT-L (поточная газификация, сухой подвод угля)	2	9	13	2	5
MCSG (поточная газификация, водоугольные смеси)	5	5	6	4	3
Остальные	17	35	44	14	18

3.3.1 Противоточные газогенераторы с подвижным слоем

В коммерческих газогенераторах с подвижным слоем уголь подводится в реактор сверху, а окислитель направляется в противотоке снизу вверх. Самая высокая концентрация кислорода наблюдается в нижней части реактора. Здесь устанавливается зона окисления, в которой происходит полная конверсия кислорода. Над ней находится зона газификации, где проходят эндотермические процессы конверсии кокса. Полученный синтез-газ направляется в верхнюю часть газогенератора и нагревает на своем пути куски угля, находящиеся в зоне осушки сырого угля и в зоне выхода летучих веществ из угля. Таким образом, утилизация тепла, выделенного при парциальном окислении угля, осуществляется непосредственно в реакторе. Благодаря этому данный тип газогенераторов отличается наименьшим удельным потреблением кислорода и наиболее низкой температурой синтез-газа на выходе из реактора (350–800 °C), по сравнению с другими технологиями газификации. По этой причине, газогенераторы с подвижным слоем на сегодняшний день являются наиболее пригодными для переработки высокозольных углей. Ввиду низкой температуры газа на выходе из реактора, полученный синтез-газ содержит значительные количества метана и высших углеводородов (смолы, масла), см. Таблицу 3. Поэтому выход синтез-газа ниже, чем с случае газификации с

Таблица 3 Типичный состав синтез-газа (об.-%, сухой газ) полученного в газогенераторе с подвижным слоем [6]

Газогенератор	CO	H_2	CO_2	CH_4	N_2, Ar, C_xH_y, …
Лурги FBDB®	20	40	27	10	3
BGL	55	28	5	7	5

кипящим слоем или поточной газификации, хотя КПД холодного газа исключительно высок. Состав синтез-газа особенно хорошо подходит для производства синтетического природного газа (substitute natural gas – SNG), но плохо подходит для производства метанола и схожих продуктов.

Зола, выходящая из газогенератора, практически не содержит остаточного углерода. Она выводится в виде сухой золы, посредством вращающихся решеток (например по технологии Лурги FBDB® или технологии SEDIN) [8] или в виде жидкого шлака, через выпускное отверстие (по технологии BGL) [9]. Все известные коммерческие газогенераторы с подвижным слоем работают с сухим подводом угля, с достаточно большим размером куска в диапазоне от 3 мм до 60 мм. Угольная пыль должна быть отделена на грохотах в процессе подготовки угля, т.к. наличие пыли приводит к нестабильному гидродинамическому режиму потока газа через подвижный слой. Следствием этого является неравномерное распределение реагирующего газа по диаметру подвижного слоя, приводящее к неполной степени конверсии угля и горячим точкам внутри подвижного слоя. Другим нежелательным эффектом нестабильного режима потока газа через слой угля являются скачки в перепаде давления внутри реактора, приводящие к выбросам газа в верхней части газогенератора и повышенному выносу золы улета из реактора.

В целом, газогенераторы с подвижным слоем отличаются низким удельным потреблением кислорода и высоким удельным потреблением пара. Затраты на подготовку угля (измельчение, отсев) относительно низкие. Рассмотренные в рамках данного исследования угли могут быть подвергнуты газификации в подвижном слое при условии, что они характеризуются относительно низкими значениями индекса свободного вспучивания. Однако состав полученного синтез-газа непригоден для производства метанола и других продуктов, ввиду сравнительно высокой концентрации метана (см. Раздел 5.2.2). По этой причине технология газификации в подвижном слое не может быть рекомендована для применения с целью глубокой переработки российских углей.

3.3.2 Газогенераторы с кипящим слоем

Коммерческие технологии газификации углей в циркулирующем кипящем слое разработаны и реализованы в промышленном масштабе различными компаниями, например Фостер Вилер, Оутотек, Энвиротерм. Эти технологии работают только при давлениях близких к атмосферному, что ограничивает их крупномасштабное применение. Технология газификации углей в стационарном кипящем слое разработана компанией ТиссенКрупп (технология HTW). Технология газификации углей в кипящем слое с высокоскоростной псевдоожижением разработана компанией Келлогг-Браун-Рут. Технология газификации углей в струйном кипящем слое разработана компанией CEC. Независимо от типа псевдоожижения, все газогенераторы с кипящим слоем работают на сухом подводе угля, с размером куска от 1 мм до 6 мм, и на сухом отводе золы. Они отличаются относительно невысоким удельным потреблением кислорода и пара. Типичные значения температуры синтез-газа на выходе из реактора лежат в диапазоне 800–1.000 °C. Ввиду средних значений температуры, газогенераторы с кипящим слоем имеют термодинамические преимущества по сравнению с поточными газогенераторами. Это приобретает особое значение в случае высокозольных углей [6], т.к. при этом отпадает необходимость в подводе тепла на расплавления золы и для ее нагрева до высокой температуры газификации, типичной для поточных газогенераторов. Поэтому газогенераторы с кипящим слоем весьма перспективны для применения в сфере газификации высокозольных углей или углей с высокой температурой плавления золы.

Однако на данный момент в эксплуатации находится только ограниченное число газогенераторов с кипящим слоем. Это прежде всего вызвано сравнительно невысокой степенью конверсии углерода, в районе 90–95%. Причиной служит неравномерное распределение времени пребывания кусков угля в реакторе, что типично для реакторов кипящего слоя. Как следствие, зола выходящая из реактора содержит 5–25 вес.-% остаточного углерода, а зола уноса даже до 50–60 вес.-% углерода. Это приводит к необходимости дожига золы в отдельном реакторе, например в расположенной поблизости угольной электростанции. Другим недостатком является то, что степень конверсии углерода в районе 90–95% [10] наблюдается только для углей с высокой реакционной способностью, прежде всего бурых углей и молодых каменных углей. Степень конверсии углерода для длиннопламенных каменных углей составляет только 60–80%. Невысокие температуры газа на выходе из реактора приводят к тому, что содержание метана составляет порядка 3–6%, что в свою очередь привело бы к снижению выхода метанола и синтетического бензина при последующей переработке синтез-газа, см. Таблицу 4. Для глубокой переработки углей шахты «Талдинская-Западная-2» и шахты «Комсомолец» не может быть рекомендована газификация по технологии кипящего слоя. Степень конверсии углерода была бы слишком низкой, что потребовало бы проведения процесса дожига золы. Это значительно ухудшило бы экономические показатели технологических цепочек, основанных на газификации в кипящем слое.

Таблица 4 Типичный состав синтез-газа (об.-%, сухой газ) полученного в газогенераторе с кипящим слоем [6]

Газогенератор	CO	H_2	CO_2	CH_4	N_2, Ar, C_xH_y, …
HTW	37	34	19	4	6

3.3.3 Поточные газогенераторы

Большинство коммерческих газогенераторов работает по технологии поточной газификации угля. Это связано с тем, что данная технология обеспечивает практически полную конверсию углерода и полученный синтез-газ отличается очень низким содержанием метана, что важно для переработки синтез-газа в химические продукты, см. Таблицу 5. Другим значительным преимуществом является высокая мощность отдельных газогенераторов. Многие производители имеют коммерческие газогенераторы мощностью 1.500–2.000 тонн угля в день. Китайский газогенератор OMB достиг производительности 3.000 тонн/день (Внутренняя Монголия, Ронгксин, 65 атмосфер, водоугольная смесь) [11]. Американский газогенератор компании Келлогг-Браун-Рут имеет производительность 5.000 тонн/день, что соответствует примерно 1.000 МВт термической мощности [12]. С другой стороны, поточная газификация требует относительно высоких затрат на подготовку угля, т.к. требуется измельчение угля до размера меньше 0,25 мм. При сухом подводе угля к тому же требуются значительные объемы газа-носителя (обычно N_2), что негативно сказывается на качестве синтез-газа. При подводе водоугольных смесей требуется большое количество воды, что снижает КПД холодного газа процесса газификации. Время пребывания в реакторе в случае поточной газификации значительно ниже (обычно 2–5 сек), чем для газификации в подвижном слое и в кипящем слое. Для того, чтобы достичь полной конверсии углерода, требуются температуры выше 1.200 °С. Температура в факеле даже выше 2.000 °С. При таких высоких температурах зола переходит в жидкое состояние. Необходимо, чтобы во всем объеме реактора температура превышала пороговое значение температуры жидкого состояния золы. В зависимости от выбора угля, эта температура может лежать в диапазоне 1.200–1.600 °С. Как правило газогенератор работает при температурах, которые на 100–150 K выше, чем температура жидкого состояния золы (из соображений безопасности) [6]. Таким образом, температура синтез-газа на выходе из реактора составляет 1.350–1.700 °С. При этих условиях жидкий шлак медленно стекает вдоль внутренних стенок газогенератора и попадает на выходе из газогенератора в водный охладитель, где происходит застывание шлака. Необходимость расплавления золы и эксплуатации газогенератора при очень высоких температурах особенно негативно сказывается на экономических показателях в случае высокозольных углей [13]. В особенности поточные газогенераторы с подводом водоугольных смесей практически не используются для переработки высокозольных углей, по термодинамическим и экономическим причинам (необходимость в нагреве воды и золы до температуры процесса газификации). Поточные газогенераторы с сухим подводом угля в принципе способны обеспечить полную конверсию углерода и особенно высокое качество синтез-газа в случае переработки углей шахты «Талдинская-Западная-2» и шахты «Комсомолец». При этом для газификации высокозольных углей шахты «Комсомолец» удельная потребность в кислороде будет выше и качество синтез-газа соответственно ниже (повышенное содержание CO_2). Все технологии поточной газификации отличаются более высоким удельным потреблением кислорода и более низким потреблением пара, чем технологии газификации в подвижном слое и в кипящем слое. Пар требуется как модератор температуры, прежде всего для углей с высокой теплотворной способностью, т.к. иначе температура процесса будет слишком высокой, что может привести к разрушению внутренних стенок газогенератора. В случае подвода водоугольных смесей, потребность в дополнительной подаче пара отпадает.

Поточныс газогенераторы различаются по конструкции внутренних стенок реактора. Газогенераторы со стенками, выложенными огнеупорным кирпичом, характеризуются более низкими капитальными вложениями но и более коротким сроком эксплуатации (от полугода до года). Термические, механические и химические нагрузки приводят к разрушению кирпичей, ввиду чего требуется их периодическая замена. Мембранные стенки более дорогостоящие, но зато они могут находится в эксплуатации в течении значительно более длительного срока (обычно производители гарантируют срок эксплуатации 10 лет, на практике сроки достигают 15–20 лет). Мембранная стенка в газогенераторе СИМЕНС (см. Рисунок 6) покрыта спиралевидными охлаждающими змеевиками, на которые нанесен огнеупорный SiC-бетон. Горячие жидкие шлаки застывают на поверхности охлаждаемой стенки и образуют твердый защитный слой, толщина которого зависит от закономерностей переноса тепла и массы [5]. Жидкий шлак стекаст вниз вдоль этого твердого слоя остывшего шлака. Таким образом стенка реактора защищается от термического воздействия и коррозии под влиянием шлаков. Мембранные стенки позволяют проводить значительно более быструю смену температуры при вводе и выводе газогенератора из эксплуатации, чем в случае реакторов со стенками, выложенных огнеупорным кирпичем. Благодаря этому сокращается время простоя и увеличивается годовой ресурс работы оборудования. С другой стороны, для газогенераторов с мембранной стенкой необходима дополнительная пилотная горелка для зажигания угольно-кислородной смеси при вводе газогенератора в эксплуатацию (см. Рисунок 6). В случае газогенераторов со стенкой, выложенной огнеупорным кирпичем, пилотная горелка не требуется, т.к. зажигание происходит самостоятельно, на раскаленной поверхности стенок.

Поточные газогенераторы могут иметь различные системы охлаждения синтез-газа. Наиболее простой, надежный и дешевый вариант – это охлаждение путем вбрызгивания воды (водный квенч). Подобные технологии предлагаются компаниями Дженерел Электрикс, Сименс, Шелл, ТиссенКрупп. Горячий газ охлаждается до температуры 180–230 °C. Шлак также попадает в водный слой, застывает и выводится из охладителя. Охлажденный и насыщенный водяным паром синтез-газ подается на очистку. Термический КПД газогенератора с водным квенчем ниже, чем для газогенераторов с котлом-утилизатором, где утилизация тепла производится в теплообменниках на базе конвекции или излучения. Котлы-утилизаторы позволяют генерировать пар высокого и среднего давления. Они улучшают КПД процесса, но одновременно

Таблица 5 Типичный состав синтез-газа (об.-%, сухой газ) полученного в поточном газогенераторе [6]

Газогенератор	CO	H_2	CO_2	CH_4	N_2, Ar, C_xH_y, …
СИМЕНС (сухой подвод угля)	60	29	5	0	6
Дженерел Электрик (водоугольная смесь)	41	34	23	0	2

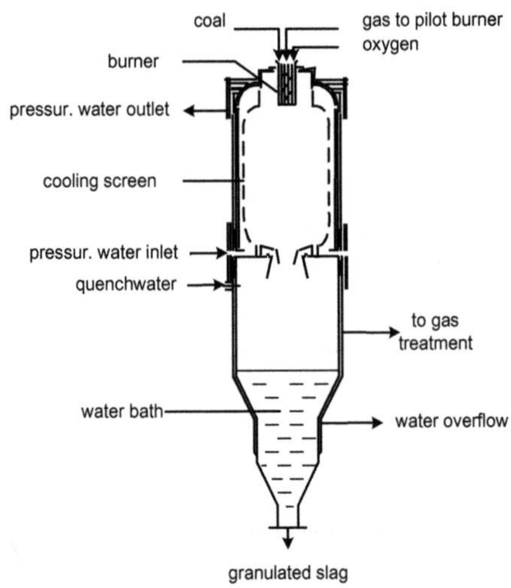

Рисунок 6 Газогенератор СИМЕНС

увеличивают необходимые капитальные вложения и сокращают годовой ресурс работы оборудования, т.к. стенки теплообменников относительно быстро подвергаются загрязнению и деструкции.

3.3.4 Инновационный газогенератор COORVED

Преимущества газогенератора с подвижным слоем, например оптимальная утилизация тепла внутри реактора и низкие капитальные вложения, могут быть комбинированы с отличительными чертами газогенератора с кипящим слоем (относительно невысокая удельная потребность в кислороде и паре) и с преимуществами поточного газогенератора (высокое качество полученного синтез-газа). С этой целью была разработана концепция многозонного газогенератора, с подвижным слоем расположенным в нижней части газогенератора и струйным кипящим слоем в верхней части реактора [14]. Соответствующая инновационная технология газификации, под названием COORVED, была успешно испытана на опытной установке во Фрайбергской горной академии (Германия). Технология COORVED характеризуется повышенными векторами скорости потока газа, направленными внутри реактора снизу вверх. На Рисунке 7 изображены четыре зоны внутри газогенератора COORVED. Зона 1 представляет собой подвижный слой в

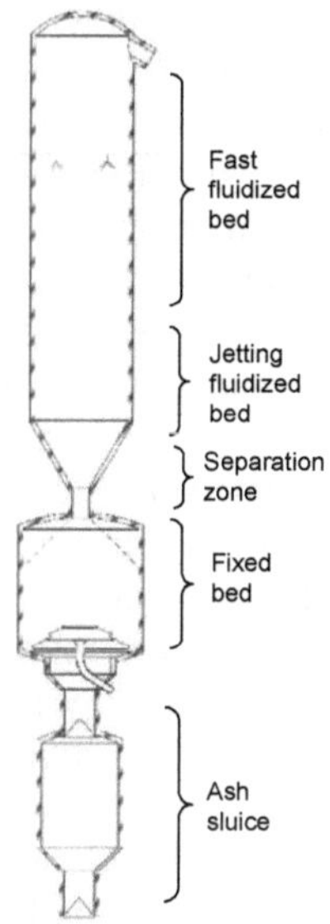

Рисунок 7 Газогенератор COORVED

нижней части газогенератора. В этой зоне агломерированные частицы золы и кокса (которые в основном образуются в струйном кипящем слое) подвергаются обработке вторичным окислителем. Окислитель представляет собой смесь кислорода и пара, которая подводится к подвижному слою через вращающуюся решетку, схожую с решеткой в технологии Лурги FBDB®. Сухая зола выводится через эту вращающуюся решетку и попадает в отводной шлюз золы. Скорость вывода золы задается скоростью вращения решетки. Над подвижным слоем расположена конусообразная разделительная зона. Здесь происходит разделение частиц по массе. Более большие куски (агломераты, куски исходного угля) движутся вниз и падают на поверхность подвижного слоя. Более мелкие частицы направляются вверх, в главную зону реакции (струйный кипящий слой и кипящий слой с высокоскоростным псевдоожижением). Лишь небольшая часть остается более длительное время в этой зоне, с образованием кипящего слоя. Данный эффект сепарации может регулироваться посредством изменения диаметра реактора. Над разделительной зоной расположен струйный кипящий слой. Это главная зона реакции в газогенераторе COORVED. Через форсунки в качестве первичного окислителя подводится кислород. Ввиду высокой скорости потока на выходе из форсунки, струя кислорода пробивает кипящий слой. В этой зоне газогенератора подаются свежие порции угля, который вступает в контакт с окислителем. Ввиду экзотермической реакции окисления угля, струйный кипящий слой отличается наиболее высокой температурой внутри реактора. В горячей зоне пламени температура лежит выше 2.000 °C. После достижения определенного уровня конверсии углерода, зола смягчается и подвергается парциальному переходу в жидкое состояние. Ввиду этого поверхность частиц золы становится липкой, что приводит к образованию агломератов. Частицы золы, ставшие слишком тяжелыми, выпадают из кипящего слоя и опускаются на дно реактора в подвижный слой. В главной зоне реакции происходит осушка кусков угля и выход летучих веществ из угля. Часть летучих веществ, прежде всего высшие углеводороды, подвергаются реакциям крекинга, ввиду относительно высокой температуры реактора. По этой причине качество полученного синтез-газа схоже с качеством газа в поточной газогенераторе. В области реактора над струйным кипящим слоем скорость потока газа наиболее высокая, так как конверсия угля сопровождается выделением газообразных продуктов. Образуется зона кипящего слоя с высокоскоростным псевдоожижением, характеристики которого лежат между циркулирующим кипящим слоем и поточной газификацией. В этой зоне больше нет остаточного кислорода, протекают лишь эндотермические реакции газификации. Температура газа опускается до 1.000–1.200 °C. При необходимости, температура газа на выходе из реактора может быть дополнительно понижена до 900–950 °C посредством вбрызгивания воды (водного квенча) в головной части реактора [14].

Вследствие относительно низкой температуры на выходе из реактора, КПД холодного газа и выход синтез-газа превышают значения, типичные для других видов газификации. Кроме того, синтез-газ отличается очень низким содержанием метана, благодаря высокой температуре (>2.000 °C) в горячей зоне пламени. Газогенератор COORVED может работать в широком диапазоне размера куска угля (0–6 мм). В связи с этим данная технология газификации пригодна для переработки угля различного размера, включая кусковой уголь и угольную пыль. Для высокозольных углей, например угля шахты «Комсомолец», технология COORVED имеет значительные преимущества по сравнению с коммерческими технологиями поточной газификации, так как отпадает необходимость полного перевода золы в жидкое состояние.

Технология COORVED позволяет существенно снизить капитальные вложения и эксплуатационные расходы. Кроме того, технология COORVED имеет преимущества по сравнению с технологиями газификации в подвижном слоем и в кипящем слое в отношении качества синтез-газа и выхода синтез-газа на тонну угля. В целом, технология COORVED позволяет значительно снизить себестоимость конечного продукта, например метанола (см. Раздел 7).

3.4. **Выбор рассматриваемых технологий газификации**

С учетом технологических и экономических аспектов, следующие технологии газификации были выбраны для оценки их применимости для глубокой переработки российских углей:

- поточная газификация СИМЕНС
- инновационная технология многостадийной газификации COORVED

Проведенный в Разделе 5 анализ возможностей переработки синтез-газа в химические продукты и синтетическое топливо показывает, что в условиях российской промышленности экономически целесообразным вариантом глубокой переработки углей представляется производство метанола, как базового продукта для дальнейшей переработки в химические продукты или в синтетическое жидкое топливо. Для производства метанола необходимо использовать, из экономических соображений, синтез-газа с минимальным содержанием метана и других углеводородов. В этой связи технология газификации в подвижном слое не может быть рекомендована для достижения поставленных целей, так как выход метана по этой технологии относительно высок, концентрация метана в синтез-газе достигает 12 об.-%. Относительно высокая температура начала деформации золы для рассмотренных углей в принципе позволяет использовать технологии газификации в кипящем слое. Однако имеется ряд негативных эффектов, понижающих экономическую целесообразность газификации данных углей в кипящем слое. Наиболее существенным является то обстоятельство, что для подобных углей степень конверсии углерода при газификации в кипящем слое недостаточно высокая (<85%).

Принимая во внимание предполагаемый спектр конечных продуктов, наиболее приемлимой коммерческой технологией для рассмотренных углей является поточная газификация. При этом для высокозольного угля шахты «Комсомолец» более подходящими являются технологии с сухим подводом угля. Преимущество может быть дано поточной технологии СИМЕНС, отличающейся значительно более низкими капитальными вложениями, чем конкурирующие технологии компаний Шелл и ТиссенКруп. Однако необходимо учитывать, что технологии поточной газификации характеризуются наиболее высокой удельной потребностью в кислороде среди всех рассмотренных технологий. С этим связана и их относительно низкая энергоэффективность (низкое значение КПД холодного газа). Особо негативную роль это играет в случае высокозольных углей, для которых глубокая переработка на базе поточной газификации как правило не находит применения в промышленном масштабе, ввиду плохих экономических показателей.

Учитывая эти обстоятельства, в данном научно-техническом исследовании была также рассмотрена применимость новой, инновационной технологии газификации COORVED, которая оптимальным образом приспособлена для глубокой переработки именно высокозольных углей. Эта технология отличается низкой удельной потребностью в кислороде, практически полной степенью конверсии углерода и высоким качеством полученного синтез-газа. Технология COORVED значительно превосходит, по техническим и экономическим показателям, существующие коммерческие технологии газификации, прежде всего в случае высокозольных углей. Однако и в применении к низкозольным углям, как например к углю шахты «Талдинская-Западная-2», технология COORVED является конкурентоспособной по сравнению с технологиями, предложенными на мировом рынке.

4. Углехимический-энергетический кластер

Для газификации и последующей переработки синтез-газа в конечный продукт имеются технически хорошо отработанные и испытанные в промышленном масштабе технологии. Основная задача на сегодняшний день связана с экономическими аспектами – снижением капитальных вложений в необходимые установки. По сравнению с технологическими цепочками, основанными на переработке нефти и газа, производство синтез-газа из угля отличается повышенными затратами на подготовку угля, очистку полученных газов, сточных вод и твердых углеродсодержащих и минеральных отходов. В этой связи, во Фрайбергской горной академии разработаны концепции, реализующие потенциал синергий между угольной энергетикой и глубокой переработкой угля. На базе таких подходов открываются новые возможности для снижения капитальных вложений в установки по производству и переработке синтез-газа и для реализации других положительных эффектов.

4.1. Описание концепции

Новый подход заключается в том, что в непосредственной близости к существующей угольной электростанции возводится сравнительно небольшая установка по глубокой переработке угля, термическая мощность которой значительно ниже установленной мощности электростанции. Данный подход получил название “Annex”. При этом рассматриваются эффекты синергии, приводящие к упрощениям для системы подготовки угля, разделения воздуха, очистки газов и синтеза. Из Annex-установки в угольную электростанцию передаются различные массовые потоки, например избыточный пар, отработанные горючие газы, сточные воды, углеродсодержащие твердые отходы. В электростанции возможна сравнительно недорогая утилизация этих потоках на существующем оборудовании. С другой стороны, на Annex-установку подаются избытки невостребованной электроэнергии, произведенной на электростанции. Общая блок-схема комбинации углехимии с угольной электростанцией показана на Рисунке 8.

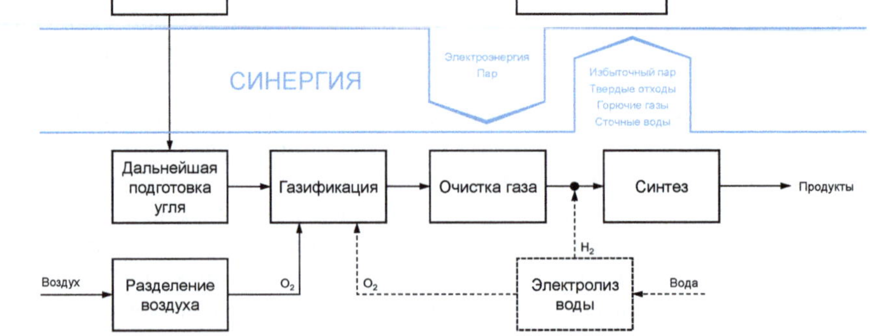

Рисунок 8 Общая блок-схема комбинации углехимии с угольной электрогенерацией

В углехимический комплекс может быть интегрирован электролизер для получения водорода и кислорода путем электролиза воды. Водород может быть подведен в процесс синтеза. Этим снижается размер необходимого реактора для процесса CO-шифт (что приводит к снижению выбросов CO_2) и увеличивается удельный выход продукта. Электричество для проведения процесса электролиза подается из угольной электростанции.

4.2. **Технические и экономические преимущества**

Интеграция технологических цепочек в угольной электростанции и в углехимическом комплексе позволяет уменьшить необходимые размеры или даже полностью исключить отдельные узлы и компоненты Annex-установки. Например отпадает необходимость очистки отходящих газов из установки Клауса, соответствующие серосодержащие газы подаются вместе с легкими горючими газами из процесса синтеза в паровой котел угольной электростанции. Полученные вредные примеси, прежде всего SO_2, улавливаются при очистке дымовых газов. Углеродсодержащие твердые отходы (например зола с остаточным содержанием углерода, образующаяся в процессе газификации с кипящим слоем) также подводятся в котел электростанции. Благодаря этому отпадает необходимость в отдельном котле-утилизаторе, дожиг золы происходит на базе существующего оборудования электростанции. Отработанная вода из газогенератора содержит большое количество сажи, ее также можно утилизировать в котле электростанции. Менее загрязненные сточные воды, например из процессов синтеза, могут быть подведены в существующие установки очистки воды при электростанции. В свою очередь, электростанция предоставляет техническую воду для парогенераторов в углехимическом комплексе. В зависимости от размера и технологии газогенератора, подготовка угля для переработки в углехимическом комплексе может быть частично или полностью реализована на существующем оборудовании угольной электростанции.

Все перечисленные аспекты приводят к снижению капитальных вложений и эксплуатационных расходов для углехимического комплекса. Проведенные ранее во Фрайбергской горной академии исследования показали, что в зависимости от мощности электростанции и газогенераторов а также используемой технологии газификации и переработки синтез-газа, может быть достигнуто снижение капитальных вложений на 15–21%. В случае создания углехимического-энергетического кластера на базе уже существующей электростанции, необходим детальный учет реальных условий эксплуатации. Не исключено, что для обеспечения надежного режима эксплуатации электростанции и углехимического комплекса необходимо заключить определенные компромиссы. Если для реализации углехимического-энергетического кластера намечена постройка новой электростанции, возможна оптимальная состыковка всех узлов оборудования и интерфейсов передачи энергетических и массовых потоков. В любом случае взаимное использование совместной инфраструктуры приводит к положительным экономическим эффектам.

Для электростанций наиболее значительные преимущества связаны с улучшенным режимом эксплуатации. Особенно ярко это проявляется в условиях европейского рынка электроэнергии. Ввиду повышения доли альтернативной энергетики, угольные электростанции вынуждены работать в очень гибком режиме, чтобы компенсировать колебания в электросети, связанные с изменчивым производством электроэнергии в солнечной и ветряной энергетике. С экономической точки зрения это приводит к уменьшению прибыли ввиду сокращения годового ресурса эксплуатации (доступности) электростанций. С технической точки зрения многократный ввод и вывод электростанций в эксплуатацию негативно сказывается на максимальном сроке эксплуатации важных компонентов электростанций, прежде всего котлов. Концепция углехимического-энергетического кластера может помочь в решении этих проблем. Благодаря углехимическому комплексу, являющемуся потребителем электроэнергии, угольная электростанция более длительное время остается в режиме эксплуатации, что положительно сказывается на экономических и технических характеристиках. Подвод горючих газов и твердых углеродсодержащих отходов в паровой котел электростанции помогает снизить потребность в угле для топки котлов.

В отличие от угольной электростанции, для которой возможен гибкий режим работы в определенном диапазоне, газогенераторы и реакторы синтеза в углехимическом комплексе как правило должны работать в непрерывном режиме. Исключением являются электролизеры, которые могут очень быстро изменять уровень потребления электроэнергии в диапазоне от 0 до 100% установленной мощности.

Проведенный во Фрайбергской горной академии, по заказу германских угольных компаний, систематический анализ показал, что для углехимического-энергетического кластера, состоящего из угольной электростанции на буром угле с электрической мощностью 600 МВт, из углехимического комплекса с термической мощностью 200 МВт (значение рассчитано на базе теплотворной способности подводимого угля) для глубокой переработки угля в различные жидкие продукты и из цеха электролизеров с общей электрической мощностью 50 МВт, моделирование технологических цепочек приводит к следующим выводам:

- Снижение себестоимости жидкого продукта примерно на 16%
- Повышение удельного выхода продукта минимум на 26%
- Снижение минимальной мощности электростанции более чем на 22%
- Снижение выбросов CO_2 в углехимическом комплексе и в угольной электростанции
- Массовые потоки из углехимического комплекса в электростанцию не оказывают существенное негативное влияние на режим эксплуатации электростанции

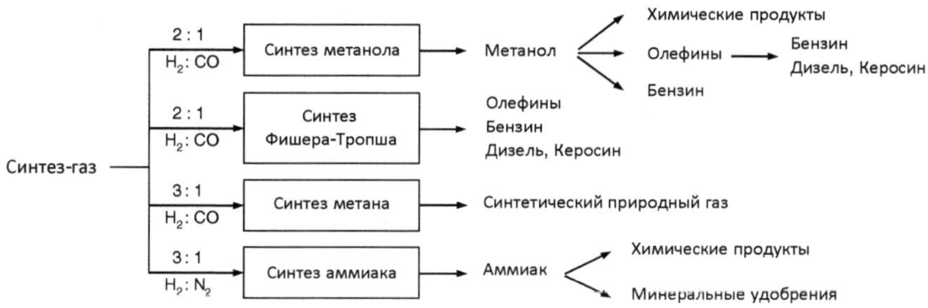

Рисунок 9 Схема использования синтез-газа

5 Технологии переработки синтез-газа

5.1. Общий обзор

Синтез-газ может быть использован для получения целого ряда продуктов, как изображено на Рисунке 9.

Одним из ключевых продуктов является метанол, который может быть получен в соответствии со следующим уравнением реакции:

$$CO + 2H_2 \rightarrow CH_3OH \tag{1}$$

Производство метанола из синтез-газа реализуется в рамках процесса, который будет более подробно описан в Разделе 5.2.2. Метанол представляет собой один из важнейших видов исходного сырья для химической промышленности, который может быть использован для получения целого спектра продуктов, например формальдегида, уксусной кислоты, этилена, пропилена. Производство этилена и пропилена, имеющих особое значение для синтеза полимеров, из метанола будет описано в Разделе 5.2.4. Другая сфера использования метанола связана с получением синтетического жидкого топлива (СЖТ). Производство СЖТ из метанола может быть описано следующим уравнением реакции:

$$CO + 2H_2 \rightarrow CH_3OH \rightarrow [-CH_2-] + H_2O \tag{2}$$

Здесь [-CH_2-] символизирует смесь углеводородов, состоящую из насыщенных, ненасыщенных и ароматических соединений. Конверсия метанола в СЖТ может быть проведена в рамках одностадийного процесса (см. Раздел 5.2.3) или в рамках двухстадийного процесса (см. Раздел 5.2.5).

Альтернативный метод использования синтез-газа представляет собой синтез Фишера-Тропша, описанный в Разделе 5.3. Механизм реакции достаточно сложен и все еще не понят в полном объеме. В упрощенном виде процесс может быть описан уравнением реакции, схожим с Уравнением (2) для производства СЖТ из метанола:

$$CO + 2H_2 \rightarrow [-CH_2-] + H_2O \quad (3)$$

Полученная смесь углеводородов включает в себя исключительно широкий спектр различных соединений, содержащих до 50 атомов углерода. Принципиальным различием между синтезом Фишера-Тропша и производством СЖТ из метанола является селективность получения продукта. Сравнение экономических параметров обоих процессов показывает, что капитальные вложения и эксплуатационные расходы для обоих процессов схожи [15]. Но это касается только производства сырого продукта. В отличие от производства СЖТ из метанола, синтез Фишера-Тропша требует применения сложной технологической цепочки переработки полученного сырого продукта в товарное топливо, существенно увеличивающей себестоимость конечной продукции.

Синтез-газ полученный при газификации угля может быть также переработан в метан:

$$CO + 3H_2 \rightarrow CH_4 + H_2O \quad (4)$$

Полученный метан называют синтетический природный газ (Substituted Natural Gas, SNG). Этот процесс описан в Разделе 5.4. Однако необходимо учесть, что затраты на получение синтетического природного газа значительно выше, чем рыночная цена на природный газ. Поэтому этот процесс экономически целесообразен только в странах с большими запасами угля и недостаточно развитой инфраструктурой для импорта природного газа.

Водород, полученный при газификации угля, может быть использован в соответствии с Уравнением (5) для производства аммиака:

$$N_2 + 3H_2 \rightarrow 2NH_3 \quad (5)$$

Аммиак представляет собой важное сырье для получения минеральных удобрений и химических продуктов (см. Раздел 5.5). Производство аммиака на базе угля в основном реализуется в Китае.

5.2. **Метанол и продукты его переработки**

5.2.1 **Общий обзор**

Метанол является одним из наиболее важных источников сырья для химической промышленности. Обзор различных сфер применения метанола приведен на Рисунке 10.

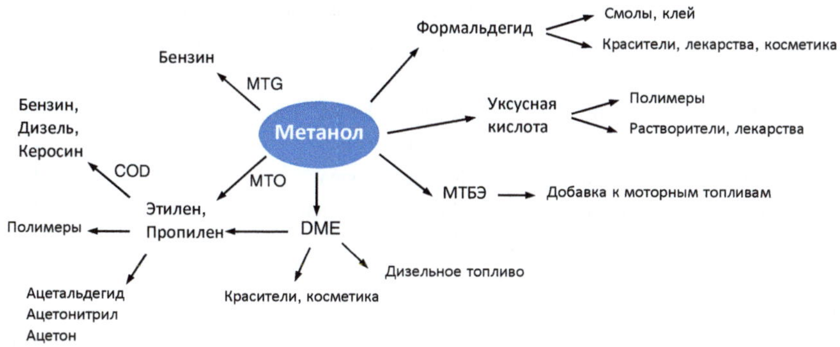

Рисунок 10 Сферы применения метанола

5.2.2 Производство метанола

Анализ рынка

Мировой объем производства метанола составляет 76 миллионов тонн/год (данные на 2016), из них более 95% производятся на базе синтез-газа [16]. Получение метанола из синтез-газа является стандартным, хорошо отработанным процессом применяемым в крупномасштабном производстве. Технологии производства метанола разработаны компаниями Лурги (27% производственных мощностей во всем мире), Джонсон Маттей/Дави (25%), Хальдор Топсэ (16%), а также Мицубиси Газ Компани, Джонсон Маттей/Уде, Джонсон Маттей/Тойо и др. [17]. Раньше в основном использовался синтез-газ, полученный конверсией природного газа. В последние годы неуклонно возрастает доля синтез-газа полученного газификацией угля, на данный момент она достигла уровня 40% [18]. Производство метанола на базе угля в основном реализуется в Китае. Ряд примеров установок по производству метанола, работающих на синтез-газе из угля, приведен в Таблице 6.

Развитие цен на метанол в Европе и Азии за последние 10 лет показано на Рисунке 11. Метанол используется для получения различных химических продуктов. Области применения метанола приведены на Рисунке 12.

Основы технологии

Производство метанола из синтез-газа в соответствии с Уравнением (1) (см. Раздел 5.1) реализуется уже на протяжении почти 100 лет. Применяемые раньше технологии при высоком давлении (250–350 атмосфер, 320–400 °C, ZnO-катализатор) были заменены, благодаря использованию новых, более эффективных катализаторов, технологиями при низких давлениях (50–100 атмосфер, 230–280 °C, $Cu/ZnO/Al_2O_3$-катализатор) [17]. Т.к. синтез метанола является равновесной реакций, однократное прохождение через реактор еще не приводит к полной конверсии синтез-газа в метанол. Максимальная

Таблица 6 Примеры установок по производству метанола на базе угля

установка	страна	год	технология	Метанол, тысяч тонн/год	Лит.
Чонгуан Дахуа	Китай	2007	Хальдор Топсэ	470	[19]
Тианжин Сода	Китай	2009	Хальдор Топсэ	470	[19]
Ксианьянг, Шаанкси	Китай	2009	Хальдор Топсэ	660	[19]
Шенхуа, Баотоу	Китай	2010	Джонсон Маттей/Дави	1.800	[20]
Чонг Тиан	Китай	2011	Лурги	3.600	[21]
Хайнан	Китай	2011	Джонсон Маттей/Дави	2.500	[22]
Датанг, Дуолин	Китай	2012	Лурги	1.800	[23]
Синопек, Ванбей	Китай	2013	Лурги	1.800	[21]
Нингбо Хейуань	Китай	2013	Джонсон Маттей/Дави	1.800	[23]
Шенхуа, Нингксиа	Китай	2014	Лурги	1.800	[23]

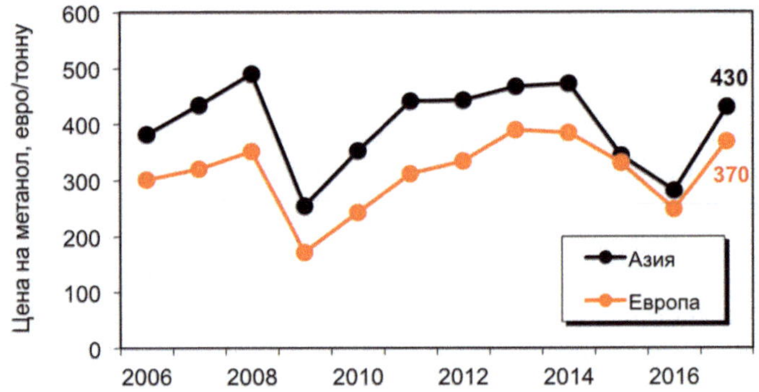

Рисунок 11 Цены на метанол в Европе и Азии [24]

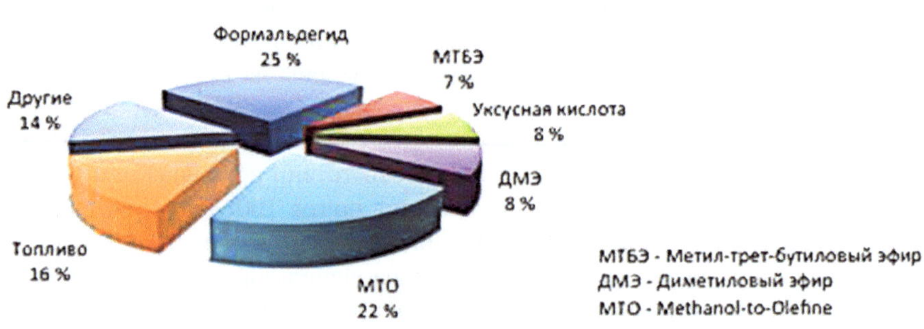

Рисунок 12 Сферы использования метанола [25]

степень конверсии синтез-газа за один проход реактора составляет около 50% [17]. Равновесие реакции смещается в сторону метанола при повышении давления или понижении температуры. Непереработанный синтез-газ рециркулируется обратно в реактор синтеза. Соотношение между рециркуляционным газом и свежим синтез-газом обычно устанавливается в диапазоне от 3 : 1 до 7 : 1 [26]. Чем выше это соотношение, тем выше конечная степень конверсии синтез-газа. Однако одновременно возрастают затраты энергии на сжатие и нагрев рециркуляционного газа. Для предотвращения повышения концентрации инертных компонентов (азот, аргон, метан и др.) часть рециркуляционного газа, обычно в пределах от 5 до 10%, выводится из кругооборота и сжигается. С термодинамической точки зрения предпочтительным продуктом реакции является метан. Его производство подавляется благодаря подбору соответствующих катализаторов. Селективность конверсии синтез-газа в метанол на коммерческих катализаторах (например от компании Джонсон-Маттей) как правило выше 99% [17].

Технологии с фиксированным слоем

Для производства метанола используются газофазные реакторы с фиксированным слоем катализатора. Самые первые технологии при низких давлениях были разработаны компанией Империал Кемикал Индустри (сегодня Джонсон-Маттей) на базе адиабатических реакторов. Эти реакторы имеют несколько слоев катализатора, помещенных на металлические решетки. Охлаждение осуществляется за счет вдувания холодного синтез-газа между слоями катализатора, поэтому КПД адиабатических реакторов ниже, чем изотермических реакторов. Преимуществом адиабатических реакторов является простота конструкции, без использования теплообменников, и соответственно более низкие капитальные вложения.

Большинство установок работают по технологии Лурги, с изотермическими трубчатыми реакторами. Синтез метанола проводится при 240–260 °C и 60–80 атмосферах, с использованием $Cu/ZnO/Al_2O_3$-катализатора (например от компании Клариант) [17, 27]. Трубки реактора заполнены катализатором. Снаружи трубки ополаскиваются кипящей водой, для отвода теплоты реакции и поддержания постоянной температуры в реакторе. Благодаря равномерному распределению температуры достигаются более высокие степени конверсии синтез-газа уже после одного прохода газа через реактор. Кроме того, утилизация пара позволяет более эффективно использовать теплоту реакции. Метанол отделяется от реакционной смеси на выходе из реактора и подвергается дальнейшей очистке в ректификационной колонне.

Для установок большей мощности (более одного миллиона тонн метанола в год) как правило используются технологии с двумя реакторами, например технология MegaMethanol® компании Лурги [28]. Эти установки отличаются более низкими удельными эксплуатационными расходами и более высокой энергоэффективностью. Технология MegaMethanol® использует один реактор

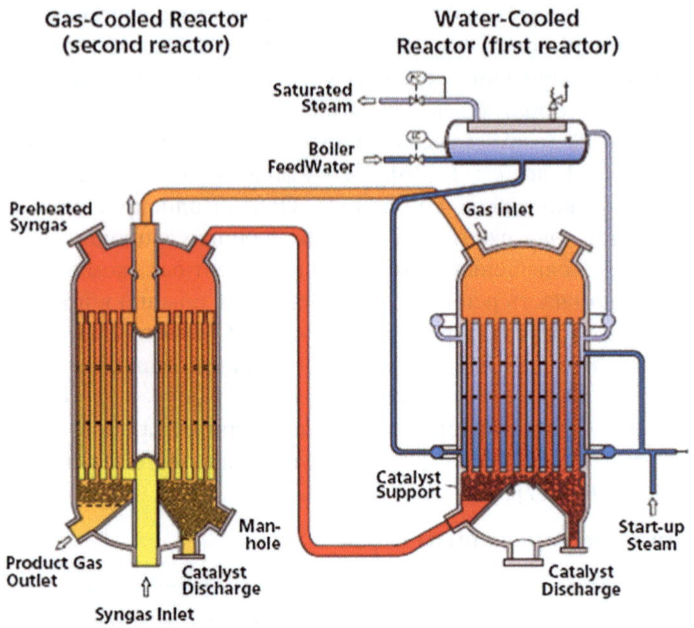

Рисунок 13 Технология MegaMethanol® компании Лурги [29]

с газовым охлаждением и один реактор с водным охлаждением, см. Рисунок 13. В реактор с газовым охлаждением подводится газ, выходящий из реактора с водяным охлаждением. Он работает при более низкой температуре, благодаря чему равновесие реакции смещается в сторону метанола. С другой стороны, для реактора с водным охлаждением средняя температура реактора и объемная скорость потока синтез-газа выше, что позволяет уменьшить объем реактора и объем используемого катализатора.

Технологии жидкофазной конверсии

Технология синтеза метанола в жидкофазном реакторе (процесс LPMeOH) была разработана компанией Айр Продактс и испытана на опытной установке компании Истман Кемикалс в Кингспорте (США) [30]. До настоящего времени технология еще не реализована в крупном масштабе. В процессе LPMeOH используется суспензия катализатора в минеральном масле. Синтез-газ проводится снизу вверх через жидкую фазу, при этом устанавливаются благоприятные условия для эффективного проведения химической реакции и отвода тепла. Жидкофазный синтез метанола отличается высокой гибкостью процесса. Соотношение H_2/CO может быть значительно ниже 2, благодаря чему возможно непосредственное использование синтез-газа из газификации угля, без предварительного изменения состава газа посредством CO-шифт

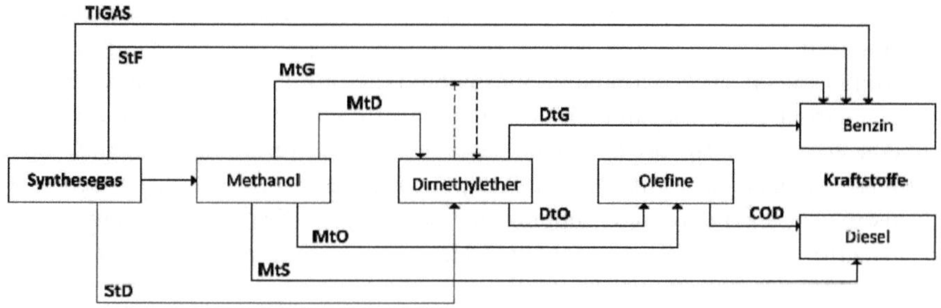

TIGAS	Topsøe Integrated Gasoline Synthesis
MtG	Methanol-to-Gasoline (z. B. ExxonMobil (Uhde))
DtG	Dimethylether-to-Gasoline (z. B. Karlsruhe Institut für Technology)
DtO	Dimethylether-to-Olefins (z. B. Karlsruhe Institut für Technology)
MtO	Methanol-to-Olefins (z. B. UOP, MtP® von Lurgi/Air Liquide)
StF	Syngas-to-Fuel (CAC, TU Freiberg)
MtD	Methanol-to-Dimethylether (z. B. Lurgi/ Air Liquide)
StD	Syngas-to-Dimethylether (z. B. JFE)
COD	Conversion of Olefins to Distillates (z. B. Lurgi/Air Liquide)
MtS	MtSynfuels® von Lurgi/Air Liquide

Рисунок 14 Обзор коммерческих технологий конверсии синтез-газа в СЖТ

процесса [31]. Недостатком по сравнению с газофазным реактором является более сложная конструкция реактора, связанная с необходимостью подготовки суспензии катализатора в минеральном масле и ее подвода в реактор.

5.2.3 Производство синтетического жидкого топлива из метанола

Метанол может быть использован для производства синтетического жидкого топлива, в рамках одностадийных или двухстадийных процессов. Обзор различных коммерческих технологий приведен на Рисунке 14. В этом разделе будут рассмотрены одностадийные технологии переработки метанола в СЖТ. По этой технологии метанол может быть переработан с высокой селективностью в синтетический бензин. Протекающие при этом стадии процесса могут быть описаны следующим образом [32]:

$$2CH_3OH \underset{+\,H_2O}{\overset{-H_2O}{\longleftrightarrow}} CH_3OCH_3 \overset{-H_2O}{\longrightarrow} \text{Leichte Olefine} \rightarrow \begin{matrix} \text{n/iso - paraffine} \\ \text{Höhere Olefine} \\ \text{Aromaten} \\ \text{Naphthene} \end{matrix} \qquad (6)$$

На первой стадии происходит дегидратация метанола с получением диметилового эфира. Диметиловый эфир находит применение в качестве замены дизеля (см. Рисунок 10). При последующей дегидратации из диметилового эфира получают легкие олефины. На последней стадии олефины реагируют в процессах переноса гидрид-иона, алкилирования и поликонденсации с получением смеси углеводородов, состоящей из высших олефинов, насыщенных углеводородов, ароматических и циклических соединений. Большая часть кислорода из молекулы метанола отводиться с водой и органическая фракция практически не содержит кислородсодержащих соединений.

Состав смеси углеводородов, образующихся при одностадийном процессе производства СЖТ, ограничен соединениями содержащими не более десяти атомов углерода. Это обусловлено механизмом протекающей реакции, более детально описаном в Разделе 5.2.5. Производство синтетического бензина с высокой селективностью процесса возможно благодаря применению катализаторов на базе цеолитов. Цеолиты представляют собой класс соединений (алюмосиликаты), имеющих каркасно-пористую кристаллическую структуру с порами, имеющими очень близкие размеры и геометрию. Благодаря равномерному распределению по размерам пор и каналов, цеолиты отличаются высокой селективностью каталитических процессов – т.н. «геометрической селективностью».

Процесс MtG

Технически наиболее зрелой технологией является конверсия метанола в СЖТ в рамках процесса „Methanol to Gasoline®" (MtG), разработанного в 70-ых годах компанией Мобил (сейчас ЭксонМобил) [33]. До 1997 года в Нью-Плимут (Новая Зеландия) эксплуатировалась установка по переработке природного газа в синтетический бензин по технологии MtG, мощностью 600 тысяч тонн бензина в год. С 2009 года в Китае работает установка по производству бензина из синтез-газа, полученного газификацией угля, мощностью 100 тысяч тонн бензина в год.

В процессе MtG метанол подвергается дегидратации с получением диметилового эфира в адиабатическом реакторе с фиксированным слоем катализатора при давлении 20–25 атмосфер и начальной температуре 300–320 ° С. Полученная равновесная смесь, состоящая из метанола, диметилового эфира и воды, подводится в реактор синтеза бензина при температуре 320–340 °С. По технологии MtG используется цеолит-катализатор марки ZSM-5. В установке в Новой Зеландии использовался адиабатический реактор. Теплота реакции отводилась из реактора через теплообменники, с генерацией пара высокого давления. Непереработанный метанол рециркулируется обратно в реактор синтеза. Полученную бензиновую фракцию подвергают перегонке и затем дальнейшей переработке, например конверсии олефинов в разветвленные алканы и удалению дурола [34]. Процесс MtG имеет целый ряд преимуществ по сравнению с синтезом Фишера-Тропша. Недостатком однако является то, что до настоящего времени не удалось реализовать одностадийный процесс с получением других видов топлива (дизель, керосин).

Процесс TIGAS

Компания Хальдор Топсе разработала технологию по переработке синтез-газа через промежуточную стадию образования метанола и диметилового эфира до получения синтетического бензина в рамках процесса TIGAS (Topsoe Integrated Gasoline Synthesis) [35]. Основной целью разработки новой технологии было снижение капитальных вложений. По технологии TIGAS конверсия синтез-газа в метанол/диметиловый эфир и последующая переработка в СЖТ производится в одном технологическом цикле, без отдельной рециркуляции непереработанного синтез-газа после метанольного реактора. Ввиду этого оба процесса проводятся при давлении 20 атмосфер. Для синтеза метанола/диметилового эфира используется разработанный Хальдор Топсе многофункциональный катализатор. Синтез бензина схож по параметрам с технологией MtG, за исключением ряда корректировок, вызванных использованием единого кругооборота рециркуляционных газов. В качстве катализатора используется цеолит ZSM-5.

Процесс STF

Еще одна технология производства синтетического бензина, под названием „Syngas to Fuel" (STF), была разработана немецкой компанией CAC и испытана на опытной установке во Фрайберге (Германия) [36]. Отличительной чертой процесса является высокое качество полученного бензина (пониженный выход ароматических соединений, высокое содержание разветвленных насыщенных углеводородов) [37]. Это прежде всего связано с особой конструкцией изотермического реактора. Кроме того используется новый цеолит-катализатор с высокой селективностью образования углеводородов со средней длиной цепочки.

5.2.4 Конверсия метанола в олефины

Анализ рынка

Мировой объем производства составляет 140 миллионов тонн/год этилена и 105 миллионов тонн/год пропилена (прогноз на 2016) [38]. В последние годы доля производства этилена и пропилена на базе метанола значительно выросла, особенно в Китае. Здесь в основном используется метанол, полученных путем газификации угля. На сегодня годовой объем производства метанола в Китае составляет 30 миллионов тонн, из них примерно 10 миллионов тонн идут на производство олефинов [38]. Ряд примеров установок по производству олефинов из метанола приведены в Таблице 7.

Развитие цен на этилен и пропилен в Европе представлено на Рисунке 15. Этилен и пропилен используются для получения различных химических продуктов, причем более 50% идут на производство полиэтилена и полипропилена.

Основы технологии

Конверсия метанола в легкие олефины описывается, как и процесс MtG, Уравнением (6), см. Раздел 5.2.3. Однако благодаря подбору соответствующих параметров процесса, конверсия может быть остановлена на стадии олефинов, которые представляют собой интермедиаты с высокой реакционной способностью на пути от метанола к синтетическому бензину [34]. При этом целесообразно рассматривать изменение прежде всего следующих параметров процесса:

Таблица 7 Примеры установок по производству олефинов из метанола

установка	страна	год	технология	Этилен/пропилен, тысяч тонн/год	Лит.
Шенхуа, Баотоу	Китай	2010	ДМТО	600	[20]
Синопек, Чонгуан	Китай	2010	С-МТО	200	[23]
Датанг, Дуолин	Китай	2012	МТР™	460	[23]
Нингбо Хейуань	Китай	2013	ДМТО	600	[23]
Wison Clean Energy	Китай	2014	УОП/Гидро	300	[23]
Шенхуа, Нингксиа	Китай	2014	МТР™	500	[23]
Yilin Energy	Китай	2014	УОП/Гидро	600	[39]
Jiutai Energy, Ордос	Китай	2015	УОП/Гидро	600	[40]
Джейанг, Йаксин	Китай	2015	УОП/Гидро	600	[23]

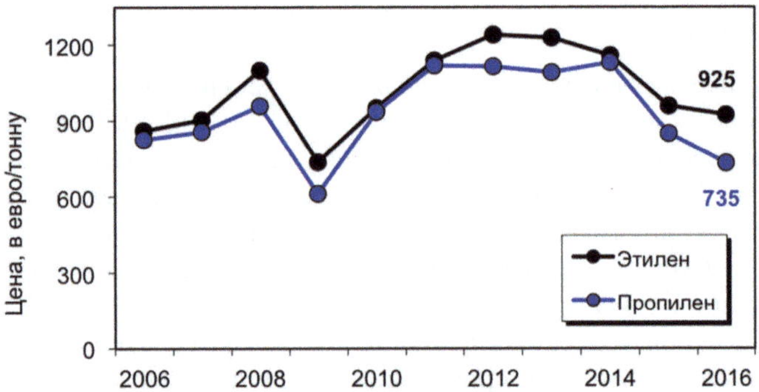

Рисунок 15 Цены на этилен и пропилен в Европе [41]

- *Уменьшение времени пребывания в реакторе и/или понижение температуры реакции*

  Недостатком при этом является, что повышение селективности получения олефинов неразрывно связано с понижением степени конверсии метанола. Ввиду этого необходима рециркуляция непереработанного метанола.
- *Уменьшение парциального давления метанола*

  При давлениях ниже одной атмосферы достигается высокая селективность образования олефинов при практически полной степени конверсии метанола.

- *Изменение SiO_2/Al_2O_3-соотношения в цеолит-катализаторе ZSM-5.* Повышение SiO_2/Al_2O_3-соотношения приводит к уменьшению кислотного характера катализатора и соответственно к понижению каталитической активности. Как следствие, повышение SiO_2/Al_2O_3-соотношения в цеолит-катализаторе ZSM-5 сопровождается повышением выхода легких олефинов и понижением доли ароматических соединений и алканов с 1–4 атомами углерода в полученной смеси углеводородов (аналогично эффекту уменьшения времени пребывания в реакторе).

В конечном итоге конверсию метанола в легкие олефины проводят с использованием цеолит-катализаторов при низких давлениях (ниже 3 атмосфер) и незначительно более высоких температурах (450–500 °C), чем в случае переработки метанола в СЖТ [42]. Полученная смесь продуктов содержит в основном бензин и легкие олефины с 2–4 атомами углерода. Предложен ряд технологий, разработанных различными компаниями.

Технология МТО компании Мобил

Компанией Мобил разработан процесс конверсии метанола в легкие олефины под названием „Methanol to Olefins“ (MTO). Конверсия метанола производится в реакторе с кипящим слоем, при температуре 470–515 °C и давлении 2,2–3,5 атмосфер [43], с использованием цеолит-катализатора H-ZSM-5 с повышеным SiO_2/Al_2O_3-соотношением [44]. Отвод тепла реакции производится посредством спиралевидных теплообменников, с генерацией водяного пара высокого давления [34]. Реактор работает в режиме непрерывной регенерации катализатора, для поддержания постоянной активности катализатора. Средняя каталитическая активность регулируется посредством изменения времени пребывания катализатора в зоне регенерации. После выхода из реактора полученная смесь продуктов разделяется в ректификационной колонне на этилен, пропилен, легкие олефины и синтетический бензин, который в основном состоит из ароматических соединений. В зависимости от заданных параметров, выход олефинов в процессе МТО изменяется в диапазоне 50–75% [45].

Технология МТР компании Лурги

Компанией Лурги разработан и реализован в крупномасштабном производстве процесс „Methanol to Propylene“ (MTP). Схема процесса изображена на Рисунке 16. На первой стадии проводится каталитическая дегидратация метанола в адиабатическом реакторе [46]. Полученная смесь подводится в адиабатический реактор, содержащий 5–6 фиксированных слоев катализатора, который работает при температуре 400–450 °C и давлении 1,3–1,6 атмосфер [47]. Используется высокоэффективный катализатор, разработанный на базе цеолит-катализатора H-ZSM-5 компанией Клариант специально для процесса МТР [48]. Данный катализатор отличается повышенной селективностью по отношению к производству пропилена, значительно пониженным выходом пропана и сниженной тенденцией к дезактивации под влиянием отложений

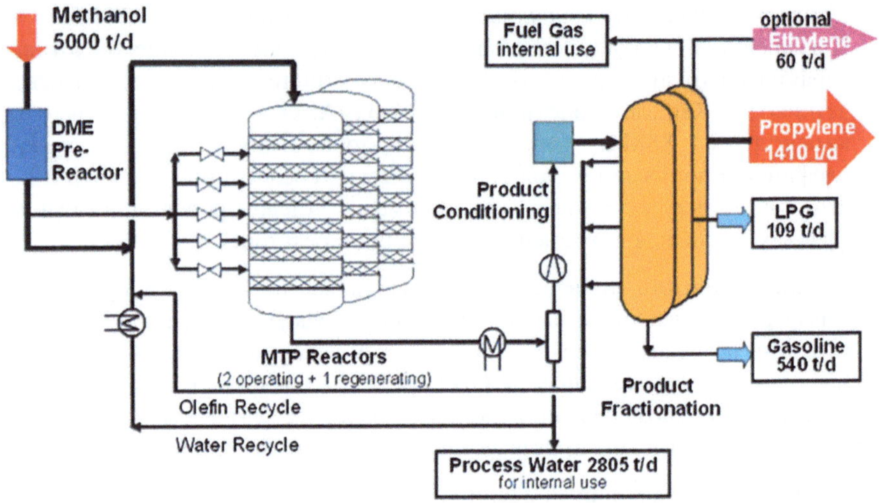

Рисунок 16 Схема процесса МТР компании Лурги [49]

кокса [48]. Как правило регенерация катализатора необходима после 500–600 часов непрерывной эксплуатации [48].

Степень конверсии смеси метанол/диметиловый эфир в процессе МТР лежит выше 99% [48]. Полученная смесь продуктов состоит в основном из пропилена, бензина и пропан-бутановой смеси, с типичным соотношением продуктов пропилен : бензин : пропан/бутан = 13 : 5 : 1 (см. Рисунок 16).

Технология МТО компании УОП/Гидро

Еще одна технология разработана американской компанией УОП и норвежской компанией Норск Гидро. Она отличается от процесса МТО компании Мобил более высокой селективностью производства этилена и пропилена, выход реакции составляет до 50% этилена и 30% пропилена [50, 51]. Реактор процесса МТО компании УОП/Гидро схож с реактором компании Мобил [50]:

- в обоих случаях применяются реакторы с кипящим слоем катализатора
- тепло реакции используется для генерации пара высокого давления

Наиболее существенным отличием является использование цеолит-катализатора SAPO-34 [51]. Этот катализатор характеризуется меньшим объемом пор, по сравнению с катализатором ZSM-5. Предполагается, что именно этим обстоятельством объясняется повышенная селективность образования легких олефинов. Путем изменения заданных параметров процесса можно регулировать в определенном диапазоне соотношение этилен/пропилен в полученном продукте [50].

5.2.5 Двухстадийные технологии производства жидкого топлива

Конверсия метанола в смесь углеводородов, из которой могут быть выделены фракции бензина и дизеля, возможна на сегодня только посредством двухстадийных технологий.

Основы технологии

На первой стадии метанол перерабатывается в смесь легких олефинов, см. Раздел 5.2.4. На второй стадии происходит процесс олигомеризации олефинов на цеолит-катализаторах. Этот процесс характеризуется следующими показателями [34]:

- Олигомеризация проводится при температурах ниже 300 °C и давлениях выше 50 атмосфер.
- Распределение продуктов определяется параметрами процесса: при высоких температурах и низких давлениях доминирует образование низкомолекулярных соединений.
- Олигомеризация сопровождается предпочтительным образованием разветвленных углеводородов, что положительно сказывается на качестве топлива.

Смесь продуктов полученных при олигомеризации олефинов разделяется в ректификационной колоне на фракции бензина, дизеля и керосина.

Технология MTO/MOGD компании Мобил

Компанией Мобил разработана двухстадийная технология, основанная на комбинации процесса MTO (см. Раздел 5.2.4) и процесса MOGD (Mobil Olefins to Gasoline/Distillate). В процессе MOGD легкие олефины образуют смесь продуктов, содержащую значительное количество тяжелых, ненасыщенных, разветвленных углеводородов, пригодных для отделения дизельной и бензиновой фракции. Общая схема комбинированного процесса MTO и MOGD изображена на Рисунке 17.

Полученная в процессе MTO смесь продуктов разделяется на этилен, легкие олефины и синтетический бензин. Легкие олефины подводятся в адиабатический реактор MOGD, вместе с частью полученного бензина. Процесс MOGD проводится при температуре 200–320 °C и давлении 30–70 атмосфер, с использованием цеолит-катализатора ZSM-5 [52, 53]. Полученный продуктов содержит большое количество разветвленных ненасыщенных углеводородов с 5–20 атомами углерода[45]. Исходный состав смеси олефинов, поступающих в реактор MOGD, оказывает лишь незначительное влияние на состав полученного продукта. Это связано с тем, что при температурах выше 230 °C устанавливается равновесие реакций олигомеризации и крекинга: при достаточно длительном времени пребывания в реакторе состав смеси продуктов определяется только температурой и давлением процесса [52]. Большим преимуществом процесса MOGD является возможность гибкого режима эксплуатации: соотношение фракций дизеля и

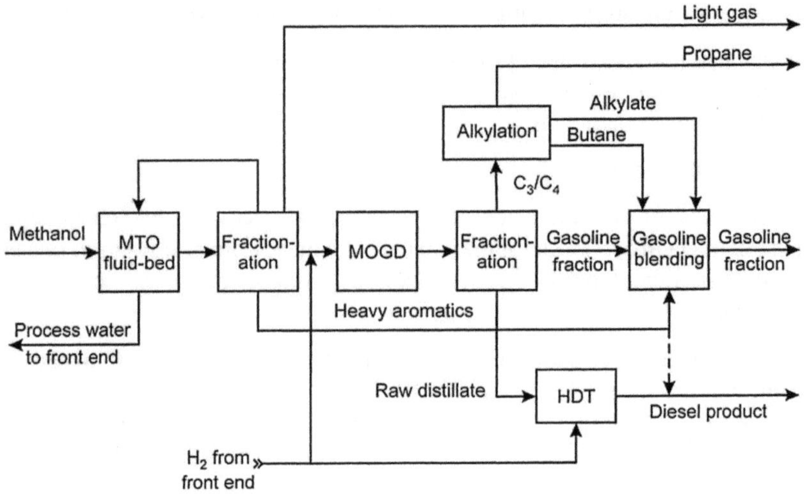

Рисунок 17 Схема комбинированного процесса МТО и MOGD [43]

Таблица 8 Распределение продуктов процесса MOGD в двух режимах эксплуатации (в качестве исходного сырья использовались C_3-C_4-олефины) [53]

фракция	Бензиновый режим	Дизельный режим	
вес.-%		Один проход	Рециркуляция[a]
C_1–C_3	4	2	2
C_4	5	4	2
Бензин (C_5 – 200°C)	84	–	–
Тяжелые фракции (>200 °C)	7	–	–
Бензин (C_5 – 165°C)	–	29	12
Тяжелые фракции (>165 °C)	–	65	84

[a]частичная рециркуляция полученного бензина

бензина может быть изменено в зависимости от параметров процесса в диапазоне от 0 до 9 [53]. Типичное распределение продуктов в бензиновом и в дизельном режиме эксплуатации показано в Таблице 8.

Соотношение дизель/бензин определяется в первую очередь следующими параметрами:

- *Температура и давление процесса MOGD*
  При температурах выше 300 °C и низких давлениях наблюдается предпочтительное образование бензина [53, 54]. Это связано с реакцией трансфера гидрид-иона, приводящей к частичной конверсии олефинов в смесь насыщенных и ароматических углеводородов.
- *Рециркуляция бензина*
  Рециркуляция полученного бензина приводит к увеличению выхода фракции дизеля/керосина, как показано в Таблице 8.

Технология MtSynfuels компании Лурги

Компанией Лурги была разработана двухстадийная технология MtSynfuels, основанная на комбинации процесса MTP (см. Раздел 5.2.4) и процесса COD (Conversion of low molecular weight Olefines to Diesel) [46]. Полученные в процессе MTP пропилен и легкие олефины конвертируются в рамках процесса COD в смесь углеводородов, пригодную для отделения бензина и дизеля. Общая схема комбинированного процесса изображена на Рисунке 18.

Разработанный компанией Лурги процесс COD используется компанией PetroSA с 1993 года для крупномасштабного производства жидких топлив в Мозель Бэй (ЮАР), мощность установки составляет 250 тысяч тонн СЖТ в год. Процесс COD проводится в адиабатических реакторах при температурах 200–280 °C и давлении 45–55 атмосфер [56, 57], с использованием цеолит-катализатора COD-9, разработанного компанией Клариант специально для процесса COD [58]. Этот катализатор применим как для реакций олигомеризации олефинов, так и для реакций крекинга и изомеризации полученных тяжелых олефинов.

5.3. Синтез Фишера-Тропша

Синтез Фишера-Тропша служит для производства синтетических жидких топлив и химических продуктов, процесс может быть описан в упрощенном виде Уравнением (3), см. Раздел 5.1. Данная технология была разработана в 30-е годы и использовалась в промышленном масштабе в Германии во время второй мировой войны и в ЮАР начиная с 50-ых годов. В последнее время новые установки Фишера-Тропша построены на Ближнем Востоке (с природным газом в качестве исходного сырья для производства синтез-газа) и в Китае (с синтез-газом из газификации угля). Обзор крупномасштабных установок Фишера-Тропша приведен в Таблице 9.

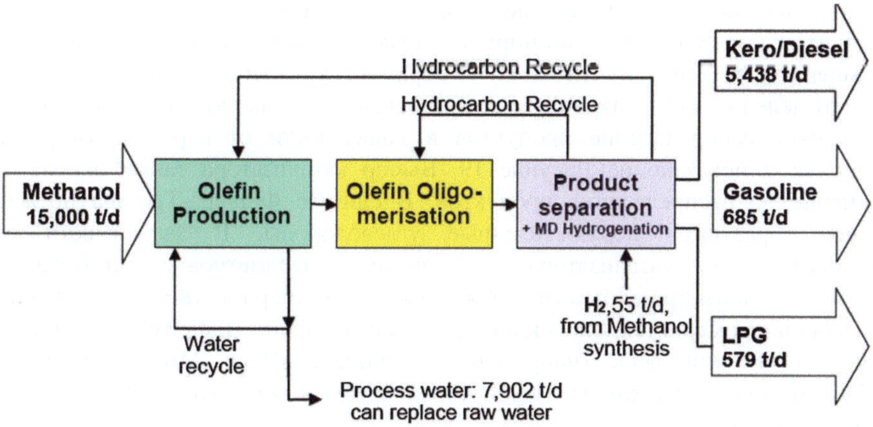

Рисунок 18 Общая схема процесса MtSynfuels компании Лурги [55]

Таблица 9 Обзор коммерческих установок Фишера-Тропша

установка	страна	год	сырье	технология	продукт тысяч тонн/ год	Лит.
Зазоль I	ЮАР	1955	уголь	ARGE	300	[59]
Зазоль II / III	ЮАР	1981	уголь	Адвансед Синтол	7.200	[60]
Джитай	Китай	2012	уголь	Синфюлс	1.000	[61]
Шанкси	Китай	2015	уголь	Синфюлс	1.000	[61]
Шенхуа	Китай	2016	уголь	Синфюлс	4.000	[61]
Моссгаз	ЮАР	1993	газ	Синтол	2.000	[60]
Бинтулу	Малайзия	1993	газ	Шелл Мидл Дистилат	660	[59]
Перл	Катар	2011	газ	Шелл Мидл Дистилат	6.300	[62]
Орикс	Катар	2006	газ	жидкофазная	1.530	[60]
Эскравос	Нигерия	2014	газ	жидкофазная	1.530	[63]

Для синтеза Фишера-Тропша разработаны высокотемпературные и низкотемпературные технологии. Для высокотемпературных процессов используются прежде всего железосодержащие катализаторы, для низкотемпературных процессов применяются как кобальт- так и железосодержащие катализаторы. Катализаторы содержат, наряду с железом и кобальтом, ряд добавок, оказывающих влияние на активность и селективность катализатора, например натрий, калий, марганец, ванадий, рутений [64]. Температура не превышает 400 °C, т.к. при более высоких температурах наблюдается быстрая дезактивация катализаторов ввиду отложений кокса на поверхности.

Отличительной чертой синтеза Фишера-Тропша является образование очень широкого спектра продуктов, содержащего различные насыщенные, ненасыщенные и ароматические углеводороды в большом диапазоне молярной массы. Соотношение продуктов зависит от температуры: при низких температурах доминируют тяжелые продукты, при высоких температурах предпочтительно образуются легкие углеводороды. Распределение по молярной массе определяется законом Шульца-Флори. Типичное распределение продуктов в зависимости от вероятности роста цепочек α показано на Рисунке 19. Выбор катализатора также оказывает влияние на распределение продуктов. В синтезе Фишера-Тропша прежде всего образуются неразветвленные углеводороды. В зависимости от используемого катализатора и других параметров эксплуатации доминирующим продуктом является бензин (например в технологии Синтол), дизель (например в технологии АРГЕ), легкие олефины (например в процессе Dow LPG) или воск (например в процессе Шелл Мидл Дистилат). Максимально достижимый выход бензина составляет около 50% и выход дизеля около 40%.

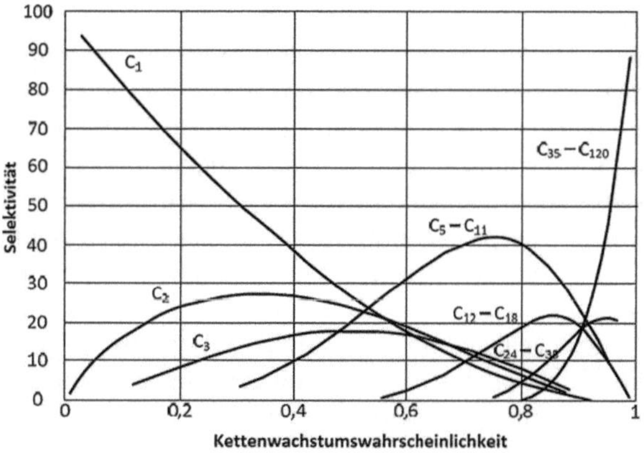

Рисунок 19 Распределение продуктов синтеза Фишера-Тропша [65]

5.3.1 Низкотемпературный синтез Фишера-Тропша

По технологии низкотемпературного синтеза Фишера-Тропша предпочтительно образуются высокомолекулярные насыщенные соединения. Они пригодны для получения дизельной фракции и воска (смазочных масел), однако требуют комплексную, многостадийную обработку. Предложен ряд коммерческих технологий, разработанных различными компаниями. Технически наиболее испытанными технологиями является процесс ARGE и жидкофазные технологии (Slurry-Bed), см. Рисунок 20.

В процессе ARGE используются изотермические трубчатые реакторы, с трубками заполненными катализатором. Реакция проводится при 220–250 °C и давлении 20–30 атмосфер, с использованием $Fe/Cu/K_2O/SiO_2$-катализатора.

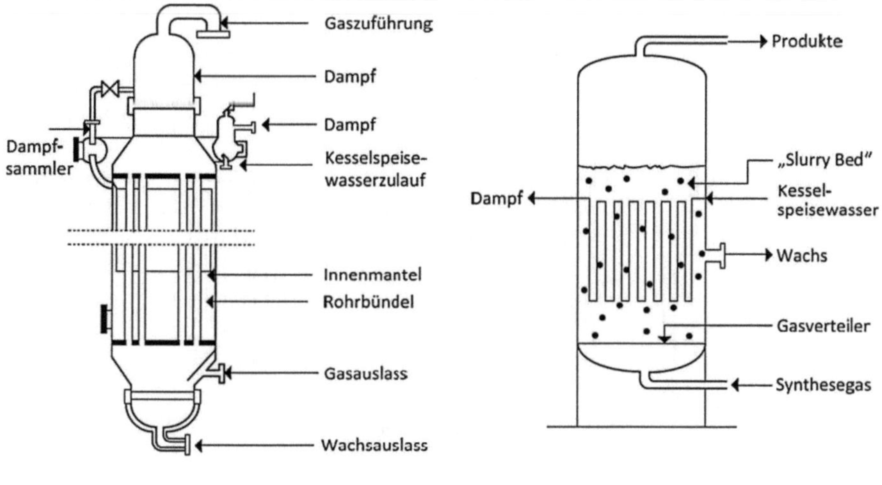

Рисунок 20 реактор процесса ARGE (слева) и жидкофазный реактор (справа) [66]

Трубки ополаскиваются кипящей водой, для отвода тепла реакции и парогенерации. Полученные высокомолекулярные продукты (воск Фишера-Тропша) стекают на дно реактора и выводятся наружу. Выход воска составляет не менее 50%, низкомолекулярные соединения (бензиновая фракция) образуются лишь в незначительных количествах.

В жидкофазном реакторе используется суспензия катализатора. Синтез-газ проводится снизу вверх, полученные высокомолекулярные продукты отводятся на дне реактора. Отвод тепла проводится через теплообменники, погруженные в жидкую фазу. Достигается более равномерное распределение температуры, по сравнению с ARGE-реактором, что способствует повышению ARGE-реактором, являются более низкие капитальные вложения и повышенная гибкость проведения процесса.

5.3.2 Высокотемпературный синтез Фишера-Тропша

Технологии высокотемпературного синтеза Фишера-Тропша используются для производства бензина и легких углеводородов. Синтез проводится в реакторах с кипящим слоем. Разработанная ранее технология Синтол была позднее вытеснена более передовой технологией Адвансед Синтол. Оба реактора представлены на Рисунке 21.

По технологии Синтол используется недорогой железосодержащий катализатор, который получают путем обработки магнетита. Синтез проводится при 300–340 °C и давлении 20–25 атмосфер. Выходящая из реактора смесь, содержащая непереработанный синтез-газ, полученные продукты реакции и катализатор, подается в сепаратор, в котором катализатор отделяется и возвращается обратно в реактор.

Усовершенствованный реактор по технологии Адвансед Синтол работает под давлением до 40 атмосфер. Используется специально разработанный $Fe/Cu/K_2CO_3$-катализатор, имеющий пониженную тенденцию к дезактивации.

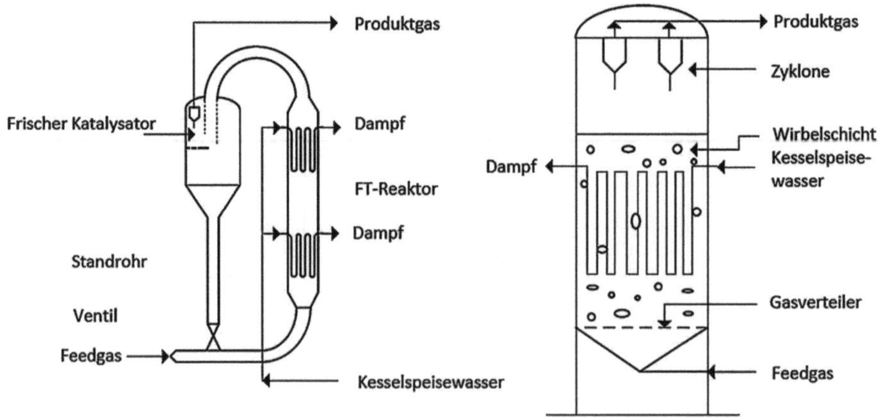

Рисунок 21 Реактор Синтол (слева) и Адвансед Синтол (справа) [66]

Высокая активность катализаторов позволяет достигать значительно более высокие степени конверсии. Процесс Адвансед Синтол может работать при меньшей концентрации катализаторов, чем в процессе Синтол, что снижает степень эрозии стенок реактора. Высокая температура реакции приводит к образованию метана, который отделяется криогенной конденсацией.

5.4. **Синтетический природный газ**

Анализ рынка

Газификация угля с последующей переработкой синтез-газа в метан является технически хорошо разработанной технологией, реализованной в промышленной масштабе. Однако необходимо заметить, что ввиду высоких капиталовложений прежде всего для установок генерации и очистки синтез-газа, синтетический природный газ не может конкурировать по себестоимости с ценами на природный газ. На сегодняшний день в мире эксплуатируется только небольшое количество крупномасштабных установок, см. Таблице 10. Большинство установок находится в Китае. Это связано со стремлением Китая к достижению независимости от импорта природного газа.

Описание технологии

Конверсия синтез-газа в метан происходит в соответствии с Уравнением 4 (см. Раздел 5.1). Как и в случае синтеза метанола и синтеза Фишера-Тропша, равновесие реакции смещается в сторону продукта при повышении давления и понижении температуры. Синтез метана проводится при 250–650 °C и

Таблица 10 Обзор коммерческих установок по производству синтетического природного газа из угля

установка	страна	год	технология	Синт. газ, милл. $м^3$/год	Лит.
Грейд Плейн Синфюлс	США	1984	Лурги	1.700	[67]
Датанг, Хексигтен	Китай	2012	Лурги	4.000	[61]
Датанг, Чифенг	Китай	2013	Джонсон Маттей/Дэви	4.000	[68]
Кинкуа, Йили Сити	Китай	2013	TREMP™	1.400	[61]
Хуиненг, Эрдос	Китай	2014	TREMP™	1.600	[61]
CPI, Йили Сити	Китай	2015	TREMP™	6.000	[68]
POSCO, Гваньянг	Южная Корея	2016	TREMP™	700	[69]
Ксинвен, Йили Сити	Китай	2016	Джонсон Маттей/Дэви	4.000	[68]
Датанг, Фуксин	Китай	2016	Лурги	4.000	[61]

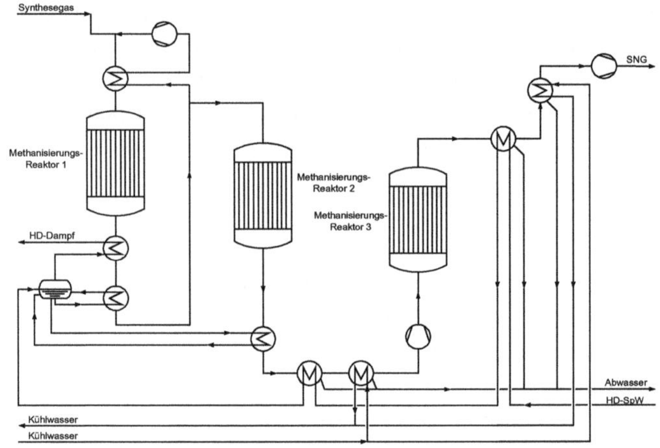

Рисунок 22 Схема процесса TREMP™ по производству синтетического природного газа [70]

давлении 50–80 атмосфер, с использованием никелевых катализаторов [70]. Ввиду высокой теплоты реакции, определяющую роль играет эффективный отвод тепла из реактора. Коммерческие технологии синтеза метана используют реакторы с фиксированным слоем катализатора или реакторы с кипящим слоем. Как правило применяются несколько адиабатических реакторов, последовательно связанных друг с другом. После выхода из реактора газовый поток подвергается охлаждению на теплообменниках. В случае реакторов с кипящим слоем достаточен один реактор. Здесь недостатком является повышенный износ частиц катализатора.

Наиболее распространенной является технология Лурги, основанная на применении двух адиабатических реакторов с фиксированным слоем катализатора. Данная технология реализована на установке Грейд Плейн Синфюлс в Северной Дакоте (США), находящейся в эксплуатации с 1984 года. Компанией Хальдор-Топсе разработана технология TREMP™ (Topsøe Recycle Energy-efficient Methanation Process), с применением трех адиабатических реакторов. Схема процесса представлена на Рисунке 22.

Технологии с применением реакторов с фиксированным слоем разработаны также компанией Джонсон Маттей/Дэви [71] и компаниями Клариант и Фостер Вил [70]. Реакторы с кипящим слоем используются в рамках процесса Синтан и процесса Би-Газ [71]. Однако оба процесса не были реализованы в установках промышленного масштаба.

5.5. **Аммиак**

Анализ рынка

Мировой объем производства аммиака составляет 200 миллионов тонн/год (данные на 2016) [72]. В последние годы растет значимость угля как сырья для

Таблица 11 Примеры установок по производству аммиака на базе угля

установка	страна	год	технология	Аммиак, тысяч тонн/год	Лит.
Анхуй Хуайнан	Китай	2009	Казале	480	[76]
Бодашиди, Вушен	Китай	2014	Келлогг-Браун-Рут	530	[77]
Хеганг, Хейлонгджинг	Китай	2014	Хальдор-Топсе	350	[77]
Хейпап Джипмей	Китай	2014	Казале	350	[77]
Джонгуан, Кушен	Китай	2014	Хальдор-Топсе	350	[77]
China Coal, Ордос	Китай	2015	Келлогг-Браун-Рут	1.000	[77]
Хенан Хунхуа	Китай	2015	Казале	650	[77]
Wulan Coal	Китай	2015	Хальдор-Топсе	750	[77]
Йангмей, Ксингксу	Китай	2015	Казале	350	[77]
Ксингьян, Уланхот	Китай	2015	Казале	350	[77]

производства аммиака. Сегодня почти 30% мирового производства аммиака получают через газификацию угля, в основном в Китае. Там производятся 65 миллионов тонн аммиака в год, из них около 80% получают на базе угля [73]. Ряд коммерческих установок по производству аммиака из угля приведен в Таблице 11. Около 85% полученного аммиака идет на производство минеральных удобрений (мочевина, сульфат аммония, нитрат аммония, фосфат аммония) [74]. Аммиак также является сырьем для химической промышленности, для производства азотной кислоты, мочевины и продуктов их переработки (полиамид, полиуретаны, полиакрилонитрил, фенол-формальдегидные смолы, меламин). Цена на аммиак лежит в диапазоне от 260 до 320 евро/тонну (в Европе, данные на 2016) [75].

Описание процесса

Синтез аммиака является хорошо отработанным процессом, который реализуется в промышленном масштабе уже на протяжении почти 100 лет без особо существенных изменений. Синтез проводится при 400–500 °C и давлении 150–250 атмосфер, с использованием железосодержащих катализаторов [74]. Равновесие реакции смещается в сторону образования аммиака при повышении давления и понижении температуры. Степень конверсии после однократного прохождения реактора обычно лежит в диапазоне 25–35%. Непереработанная смесь H_2/N_2 отделяется от аммиака посредством криогенной конденсации при температурах ниже 0 °C и направляется обратно в реактор.

В промышленном масштабе синтез аммиака проводится в реакторах с фиксированным слоем катализатора. Как и в других процессах переработки синтез-газа, определяющую роль играет эффективный отвод тепла из реактора. Для этого используются либо изотермические трубчатые реакторы либо адиабатические реакторы. Примеры обоих типов реактора приведены на Рисунке 23. В первом случае трубки теплообменников проводятся через слой катализатора. Внутри трубок пропускается холодный синтез-газ, для обеспечения отвода тепла и одновременного нагрева синтез-газа до пороговой температуры начала реакции. Коммерческие технологии разработаны компаниями ИСИ, НЕК, Казале и др. Большинство современных аммиачных установок используют адиабатические реакторы с несколькими слоями катализатора. Преимуществом таких реакторов является простота конструкции и эффективность отвода тепла. Для охлаждения реактора вводится холодный синтез-газ между слоями катализатора (прямое охлаждение) или же используются теплообменники (непрямое охлаждение), с генерацией пара высокого давления. Реакторы с непрямым охлаждением отличаются более высоким КПД и обычно используют в установках большой мощности. Коммерческие технологии разработаны компаниями Удэ, Казале, Келлогг, Хальдор Топсе.

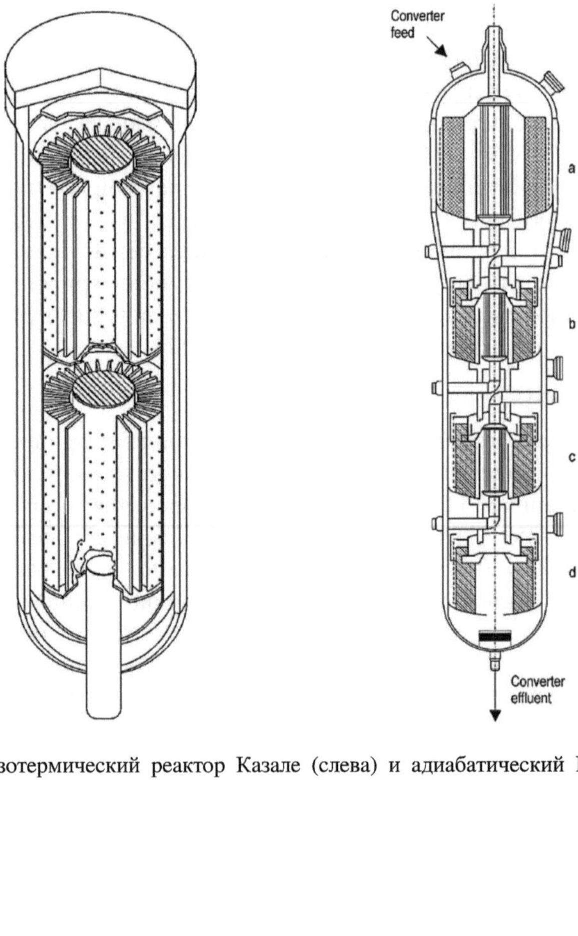

Рисунок 23 Изотермический реактор Казале (слева) и адиабатический Келлогг-реактор (справа) [74]

5.6. **Выбор рассматриваемых продуктов синтеза**

С учетом технологических и экономических аспектов, для оценки применимости процесса газификации для глубокой переработки угля в России были выбраны следующие продукты синтеза:

- метанол
- синтетический бензин, полученный из метанола посредством процесса MtG

Метанол является одним из важнейших видов сырья для химической промышленности, с годовым приростом мировых объемов производства 6% [16]. Подавляющая доля метанола производится из синтез-газа. При этом неуклонно растет значение синтез-газа, получаемого газификацией угля, на сегодня его доля достигла уровня 40%. В зависимости от региональных условий и национальной экономической политики, использование угля в качестве сырья для производства метанола вполне может быть конкурентоспособным по сравнению с производством на базе природного газа. Цена метанола на мировом рынке лежит в диапазоне 250–500 евро/тонну, при этом развитие мировой экономики оказывает более существенное влияние на ценообразование для метанола, чем для химических продуктов получаемых из метанола (например этилен или пропилен).

В качестве примера для переработки метанола в химические продукты в данном исследовании было рассмотрено получение этилена и пропилена из метанола по технологии „Methanol to Olefine“. Потребность в этилене и пропилене в химической промышленности непрерывно растет, прежде всего для производства полимеров (годовой прирост мирового объема производства составляет для этилена 4–5%, для пропилена 5–6%) [78]. При оценки применимости российских углей для глубокой переработки эти продукты синтеза были исключены из рассмотрения. Ввиду того, что в этом случае необходим химический кластер, соединяющий предприятия углехимии с производством и обработкой полимеров, для дальнейшей переработки газообразных продуктов этилен и пропилен (например с получением полиэтилена и полипропилена) в непосредственной близости от их производства. Это возможно в регионах с хорошо развитой химической промышленностью. В качестве альтернативы может быть рассмотрена поставка метанола для китайского рынка, что представляется целесообразным в первую очередь для глубокой переработки углей на угольных месторождениях Кузбасса. В Китае прирост мощностей по переработке метанола в олефины опережает рост мощностей по производству метанола из угля. В этой связи уже сейчас ряд установок по производству олефинов, выведенных в эксплуатацию в последние годы, ориентируются на закупку метанола на мировых рынках.

Другая возможность, рассмотренная в данном исследовании, представляет собой переработку метанола в синтетическое жидкое топливо. Здесь возможно использование одностадийных технологий (например MtG) или двухстадийных технологий (например MtSynfuels®). В отличие от одностадийных технологий, позволяющих получать только бензин, в двухстадийных технологиях возможно получение фракций бензина, дизеля и керосина. Но они требуют более сложных установок, что негативно сказывается на экономических параметрах. Поэтому в данном исследовании рассматривается технологическая цепочка получения бензина одностадийным процессом MtG.

Востребованным вариантом для глубокой переработки российских углей могло бы быть получение аммиака. Этот вариант был пока исключен из рассмотрения, так как и в этом случае целесообразно внедрение химических кластеров (включающих производство минеральных удобрений) для непосредственой переработки полученного газообразного аммиака.

6. **Избранные технологические цепочки**

На основе анализа с учетом технических и экономических параметров, проведенного в Разделе 2 (Выбор углей), Разделе 3 (Технологии газификации) и Разделе 5 (Технологии переработки синтез-газа), были выбраны две технологии газификации и два продукта переработки синтез-газа. Это поточная газификация по технологии СИМЕНС и новая технология газификации COORVED, основанная на комбинации технологии циркулирующего кипящего слоя, парциальной агломерации зольных частиц в верхней части реактора и вторичной зоны газификации в подвижном слое в нижней части реактора. Среди продуктов переработки синтез-газа выбраны метанол и синтетический бензин, полученный из метанола процессом MtG. На Рисунке 24 даны обозначения для всех восьми рассмотренных цепочек.

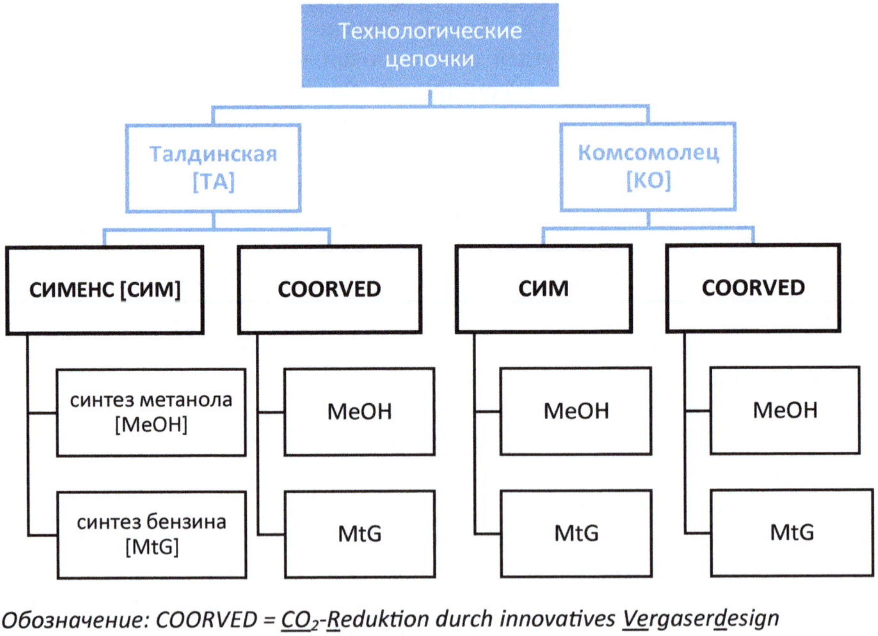

Рисунок 24 Обзор рассмотренных в исследовании технологических цепочек

1. цепочка: ТА-СИМ-MeOH	5. цепочка: КО-СИМ-MeOH
2. цепочка: ТА-СИМ-MtG	6. цепочка: КО-СИМ-MtG
3. цепочка: ТА-COORVED-MeOH	7. цепочка: КО-COORVED-MeOH
4. цепочка: ТА-COORVED-MtG	8. цепочка: КО-COORVED-MtG

В этом разделе будут более подробно описаны все рассмотренные технологические цепочки. Кроме того будут представлены наиболее важные рамочные условия, которые были учтены при оптимизации выбранных технологических цепочек при помощи коммерческого программного пакета Aspen Plus®.

6.1. **Газогенератор**

Для обоих технологий газификации рядовой уголь вначале подвергается измельчению и отсеву заданной фракции. В технологии COORVED применяется фракция 0–6 мм. Для поточной газификации размер частиц не должен превышать 200 микрон, т.к. в газогенератор вдувается угольная пыль. Подвод угля в поточный газогенератор осуществляется посредством смешения угольной пыли с высококонцентрированным CO_2, полученном в процессе очистки синтез-газа. Подвод угля в газогенератор COORVED происходит самостоятельно (частицы угля падают в газогенератор).

Производство кислорода, необходимого для процесса газификации, проводится в установках по разделению воздуха. Они работают по принципу криогенного сжижения и ректификации воздуха. Схема процесса приведена на Рисунке 25.

При моделировании заданная чистота кислорода составляла 99 об.-%. Высокая чистота кислорода необходима для ограничения содержания

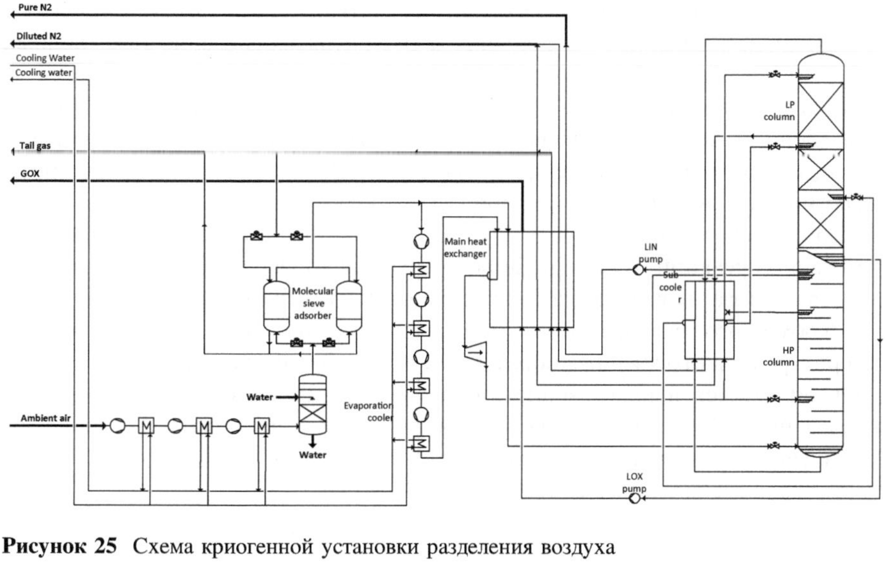

Рисунок 25 Схема криогенной установки разделения воздуха

Таблица 12 Рамочные условия, учтенные при моделировании установки по разделению воздуха

показатель	ед. изм.	значение
Температура на выходе после охлаждения сжатого воздуха	°C	30
Температура на выходе после испарителя	°C	20
Давление после декомпрессии жидкого воздуха	атмосфер	14
Температура выхода полученных газов (кислород, азот)	°C	20
Давление вверху ректификационной колоны высокого давления	атмосфер	5,12
Давление вверху ректификационной колоны низкого давления	атмосфер	1,31
Разница давления в колоне высокого давления	миллибар	175
Разница давления в колоне низкого давления	миллибар	40
Число теоретических тарелок в колоне высокого давления		35
Число теоретических тарелок в колоне низкого давления		25

инертных компонентов в полученном синтез-газе. Повышенная концентрация инертных компонентов привела бы к тому, что в процессе синтеза наблюдалось бы их дальнейшее накопление в непереработанном синтез-газе, который рециркулируется обратно в реактор. Это бы в свою очередь привело бы к повышению энергозатрат компрессоров (как для свежего синтез-газа, так и для рециркулирующего синтез-газа). Ряд других рамочных условий, учтенных при моделировании, приведен в Таблице 12.

6.1.1 **Поточная газификация**

В поточной газификации по технологии СИМЕНС в реактор интегрирован охладительный экран для генерации пара среднего давления. Синтез-газ выходит из газогенератора с высокой скоростью потока (около 25 м/сек) и при высокой температуре и попадает в резервуар охлаждения газа (водный квенч), расположенный непосредственно под газогенератором. Подогретая вода впрыскивается прямо в поток горячего газа и вызывает быстрое охлаждение газа до равновесной температуры водо-паровой смеси, которая определяется общим давлением процесса. Из производственного опыт известно, что связанное с этим процессом повышение концентрации пара в газовой смеси положительно сказывается на смещении равновесия гомогенной реакции CO-шифт. Дополнительная конверсия CO, по сравнению с составом газа на выходе из горячей зоны газогенератора, составляет около 3%. Кроме того, расплавленные минеральные компоненты угля застывают в резервуаре водного квенча с образованием твердого шлака. Гранулированный шлак опускается в шлам водной фазы и выводится вместе с ней из зоны охлаждения. Синтез-газ, насыщенный водяным паром, выходит из зоны охлаждения при температуре около 200–250 °C. На Рисунке 6 (см. Раздел 3.3.3) изображена схема газогенератора СИМЕНС.

Таблица 13 Рамочные условия при моделировании газогенератора СИМЕНС

показатель	ед. изм.	значение
Давление газогенератора	атмосфер	40
Температура газификации для угля шахты «Талдинская-Западная-2»	°C	1.650
Температура газификации для угля шахты «Комсомолец»	°C	1.550
Плотность смеси при подводе угольной пыли в потоке CO_2 в газогенератор	кг/м^3	350
Температура окислителя перед входом в газогенератор	°C	240
Давление водяного пара в охладительном экране	атмосфер	47

При моделировании газогенератора учитывался описанный эффект дополнительной конверсии CO. Температура газификации выбрана таким образом, что она лежит примерно 150 °C выше температуры жидкого состояния золы, чтобы гарантировать полный переход золы в жидкое состояние. По этой причине выбранная температура газификации для углей ТА и КО различается (разница между обоими температурами составляет 100 °C). Некоторые из рамочных условий, учтенных при моделировании, приведены в Таблице 13.

6.1.2 Газогенератор COORVED

Подробное описание технологии COORVED приведено в Разделе 3.3.4. При моделировании учтено, что степень конверсии углерода по этой технологии близка к 100%, т.е. полученная зола практически не содержит остаточный углерод. Поэтому отпадает необходимость в проведении сложного процесса дожигания золы. Этим технология COORVED отличается от существующих коммерческих технологий, например HTW. Синтез-газ выводится из газогенератора в верхней части реактора и попадает в резервуар конвекционного охлаждения, в котором охлаждение газа сопровождается генерацией пара среднего и высокого давления. Схема газогенератора COORVED изображена на Рисунке 7 (см. Раздел 3.3.4).

В качестве первичного окислителя используется паро-кислородная смесь с высоким содержанием кислорода. Подвод окислителя приводит к установлению высоких температур в зоне первичной газификации. Эти температуры значительно превышают температуру размягчения рассматриваемых углей (1.440 °C в случае угля шахты «Талдинская-Западная-2» и 1.310 °C для угля шахты «Комсомолец»), в следствии чего наблюдается парциальная агломерация частиц золы. Полученные агломераты, содержащие значительное количество остаточного углерода, опускаются на дно реактора после того, как они достигли определенного критического размера. В нижней зоне реактора образуется подвижный слой частиц золы. Вторичный окислитель, представляющий собой паро-кислородную смесь с плотностью 7 кг/м^3 (н.у.), вдувается снизу через решетку. Благодаря этому дополнительному окислителю достигается полная конверсия остаточного

Таблица 14 Рамочные условия при моделировании газогенератора COORVED

показатель	ед. изм.	значение
Давление газогенератора	атмосфер	40
Температура газа перед охлаждением в реакторе	°C	1.200
Температура газа на выходе из реактора	°C	950
Давление пара в охладителе (режим высокого давления)	атмосфер	140
Давление пара в охладителе (режим среднего давления)	атмосфер	40

углерода в частицах золы. В Таблице 14 представлены наиболее важные рамочные условия, учтенные при моделировании газогенератора. Температура газа перед охлаждением составляет для обоих углей 1.200 °C. В верхней части реактора впрыскивается вода для частичного охлаждения газа до 950 °C, после выхода из реактора синтез-газ подается в конвекционный охладитель.

6.2. **Очистка синтез-газа**

После охлаждения происходит очистка и подготовка синтез-газа, включающая три последовательных процесса: водная промывка газа, процесс CO-шифт и промывка газа для удаления кислых компонентов (прежде всего сероводород и CO_2). Сероводород подвергается в установке Клауса переработке в элементарную серу.

6.2.1 **Водная промывка газа**

Процесс водной промывки газа производится под давлением. При этом удаляются остатки золы улета а также аммиак и легколетучие хлориды. Для упрощения, моделирование производится на базе рассмотрения четырехступенчатой абсорбционной колоны. При этом учитываются константы диссоциации поглощенных соединений. Большая часть промывной воды рециркулируется в рамках закрытого кругооборота. Для предотвращения повышения концентрации растворенных примесей (например для хлоридов максимально допустимое содержание составляет 7.000 ppm) часть воды, обычно в пределах до 10%, выводится из кругооборота. В Таблице 15 собраны наиболее важные параметры моделирования.

Таблица 15 Рамочные условия при моделировании водной промывки синтез-газа

показатель	ед. изм.	значение
Число теоретических ступеней		4
Давление на входе	атмосфер	38
Потеря давления	атмосфер	1

6.2.2 Процесс CO-шифт

После водной промывки устанавливается, посредством процесса CO-шифт, необходимое соотношение концентраций основных компонентов синтез-газа (т.н. модуль синтез-газа). Для этого поток синтез-газа разделяется на две части, одна из которых идет в реактор CO-шифт, а другая посылается в обход реактора. Для процесса CO-шифт используются два адиабатических реактора с фиксированным слоем катализатора на базе сульфида кобальта и молибдена. Отличительной чертой этого катализатора является то, что он может быть применен и для газов с высоким содержанием сернистых примесей. В случае относительно сухого синтез-газа из газогенератора COORVED (парциальное давление водяного пара значительно ниже равновесного значения) необходима добавка дополнительного количества пара, для этого используется водяной пар среднего давления. Отвод тепла производится путем генерации пара. Температура на входе в реактор составляет 280 °C. На выходе температура не должна превышать 490 °C, чтобы исключить негативное воздействие на стабильность катализатора. На Рисунке 26 изображена схема процесса CO-шифт, включая процесс водной промывки газа. В Таблице 16 собраны наиболее важные параметры моделирования процесса CO-шифт.

6.2.3 Промывка газа для удаления кислых примесей

После процесса CO-шифт синтез-газ подвергается охлаждению и отделению конденсата. Затем газ подается на промывку для удаления серосодержащих соединений, CO_2 и остаточного содержания других примесей (NH_3, хлориды). Этот процесс моделируется в соответствии с технологией Ректизол, использующей метанол для селективного удаления сероводорода и CO_2. Данная технология отличается от конкурирующих процессов (например Селексол) более высокой степенью очистки газа и повышенной селективностью удаления H_2S и CO_2. Во всех рассмотренных случаях заданная чистота очищенного газа по отношению к остаточной концентрации

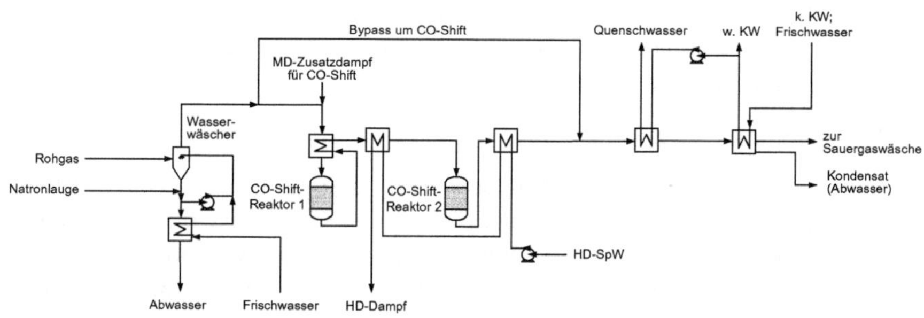

Рисунок 26 Схема водяной промывки газа и двухступенчатого адиабатического процесса CO-шифт

Таблица 16 Рамочные условия при моделировании процесса CO-шифт

показатель	ед. изм.	значение
Соотношение пар/сухой газ	кмоль/кмоль	>1,5
Соотношение H_2S/H_2O	кмоль/кмоль	>0,001
Температура на входе в реактор	°C	280
Температура на выходе из реактора (первый реактор)	°C	<490

Таблица 17 Требования к качеству синтез-газа для синтеза метанола

компонент	ед. изм.	значение
CO	об.-%	30–32
H_2	об.-%	64–68
CO_2	об.-%	соответств.
$CH_4 + N_2 + Ar$	об.-%	<4
Ar	об.-%	нет
H_2O	–	инертн.
сера (общее содержание)	мг/м^3 (н.у.)	<0,1
щелочные металлы	мг/м^3 (н.у.)	<0,2
пыль	мг/м^3 (н.у.)	<0,2

сероводорода была ниже 0,1 ppmv. Это соответствует требованиям к чистоте синтез-газа для синтеза метанола (см Таблицу 17).

Процесс промывки газа проводится при температурах -50 °C и ниже. Используются две абсорбционные колоны: первая для удаления H_2S, вторая для удаления CO_2. Для охлаждения используется аммиачная холодильная установка. При моделировании были учтены возможности тепловой интеграции процессов, для максимально возможного снижения относительно высоких затрат энергии на эксплуатацию холодильной установки. Регенерация метанола после насыщения H_2S и CO_2 производится в две стадии, на первой этапе происходит понижение давления, на втором этапе разгонка в ректификационной колоне. Полученный после регенерации отходящий газ с высоким содержанием H_2S направляется в установку Клауса. Схема процесса промывки газа показана на Рисунке 27. Главные параметры моделирования представлены в Таблице 18.

6.2.4 Установка Клауса

В установке Клауса происходит переработка H_2S в твердую серу. Имеется целый ряд коммерческих технологий для процесса Клауса, отличающихся выбором окислителя, методом очистки отработанных газов и др. При моделировании в рамках данного исследования в качестве окислителя рассматривался обогащенный кислородом воздух. Отработанный газ из процесса Клауса примешивается к синтез-газу, идущему на промывку для

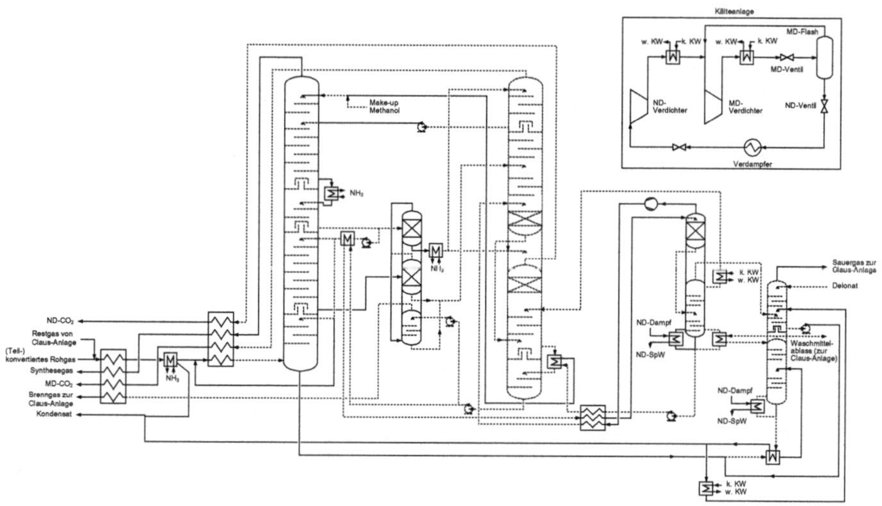

Рисунок 27 Схема промывки газа для селективного удаления кислых компонентов

Таблица 18 Рамочные условия при моделировании процесса удаления кислых компонентов

показатель	ед. изм.	значение
Температура промывки	°C	- 40
Содержание CO_2 в очищенном газе	об.-%	<3,0
Содержание H_2S в очищенном газе	ppmv	<0,1
Число теоретических ступеней в абсорбционной колоне H_2S		5
Число теоретических ступеней в абсорбционной колоне CO_2		18
Температура в верхней части/нижней части колоны регенерации	°C	30/95

удаления кислых примесей. Благодаря этому удается упростить конструкцию используемой установки Клауса.

Газ, содержащий H_2S, на первой стадии направляется в камеру парциального сгорания. Использование обогащенного кислородом воздуха позволяет улучшить степень очистки газа и ускорить протекающие процессы. Поток газа, выходящий из камеры сгорания, уже содержит некоторое количество элементарной серы. Этот поток газа подается в теплообменник для отвода тепла реакции, полученный при этом пар среднего и низкого давления используется для различных процессов подогрева реакционной смеси. Затем происходит конденсация серы в барабанах с водным охлаждением, при температуре около 70 °C. Полученная элементарная сера (модификация S_8) представляет собой товарный продукт, направляемый потребителям. Непереработанный газ после камеры сгорания направляется в два последовательно расположенных реактора с фиксированным слоем

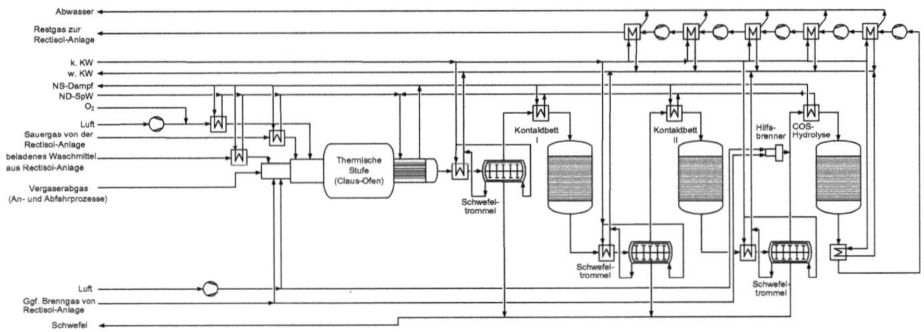

Рисунок 28 Схема установки Клауса

Таблица 19 Рамочные условия при моделировании установки Клауса

показатель	ед. изм.	значение
Степень O_2-обогащения воздуха применяемого в установке Клауса	об.-%	40
Температура газа перед подводом в слой катализатора	°C	200
Температура газа перед подводом в реактор гидролиза COS	°C	150

катализатора. Между реакторами, работающими в адиабатическом режиме, также производится конденсация образовавшейся серы в барабанах-охладителях. Оставшийся непрореагировавший газ содержит незначительное количество органических соединений серы (COS, CS_2). Поэтому он подается в реактор гидролиза, для конверсии серосодержащих соединений в H_2S. Затем газ направляется в компрессор и после этого примешивается к синтез-газу, идущему на кислую промывку. Схема установки Клауса показана на Рисунке 28. В Таблице 19 представлен ряд параметров моделирования установки Клауса.

6.3. Синтез метанола

Синтез-газ, очищенный от вредных примесей вызывающих дезактивацию катализаторов, подается в компрессор и затем направляется в реактор синтеза метанола. Для синтеза метанола предложен целый ряд коммерческих технологий, использующих газофазные реакторы (адиабатические или изотермические) или жидкофазные реакторы. Обзор технологий проведен в Разделе 5.2.2. Для моделирования была выбрана технология Лурги с изотермическим трубчатым реактором. Так как мощность рассмотренной установки относительно небольшая, была выбрана технология с одним реактором. Для установок большей мощности как правило используются технологии с двумя реакторами, например MegaMethanol® компании Лурги (см. Раздел 5.2.2). Оптимальная температура реакции в диапазоне 240–260 °C

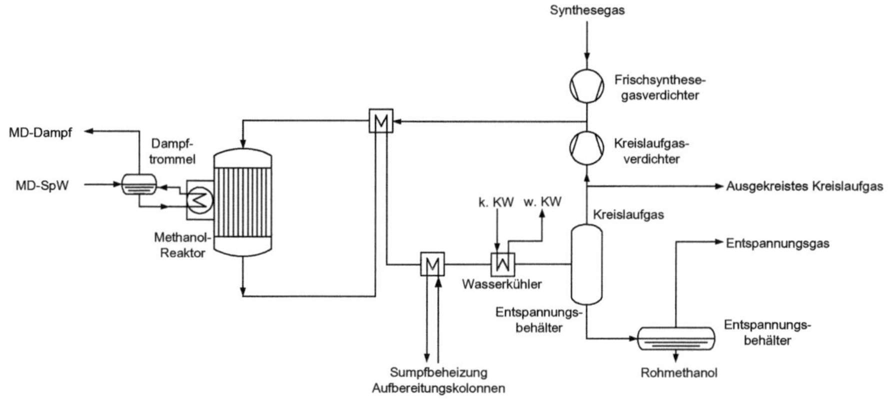

Рисунок 29 Схема синтеза метанола

поддерживается путем ополаскивания трубок реактора кипящей водой (генерация пара среднего давления). Селективность конверсии синтез-газа в метанол составляет не менее 99%. Состав синтез-газа должен соответствовать заданным параметрам (модуль синтез-газа в диапазоне 2,05–2,10 и концентрация CO_2 не менее 2,5 об.-%). Схема синтеза метанола изображена на Рисунке 29. В Таблице 20 представлены наиболее важные параметры моделирования синтеза метанола.

Продукт реакции после синтеза метанола содержит воду и другие компоненты и поэтому должен быть подвергнут очистке в ректификационной колоне, с отделением легких газов (CO, CO_2, H_2, CH_4, N_2, и др.), диметилового эфира, кетонов, формиатов, альдегидов, воды и высококипящих примесей, например тяжелых углеводородов, органических кислот, высших спиртов. В промышленном масштабе очистку метанола проводят с использованием двух или трех ректификационных колон. В первой колоне происходит отделение легколетучих компонентов. Полученный сырой метанол направляют в ректификационную колону

Таблица 20 Рамочные условия при моделировании синтеза метанола

показатель	ед. изм.	значение
Соотношение рециркуляционный газ / свежий газ	кмоль/ кмоль	3,5
Температура на выходе из реактора	°C	255
Давление реактора	атмосфер	60
Потеря давления в реактора	атмосфер	1
Модуль синтез-газа перед входом в реактор	кмоль/ кмоль	2,07
Давление пара генерируемого в охладителе (режим среднего давления)	атмосфер	40

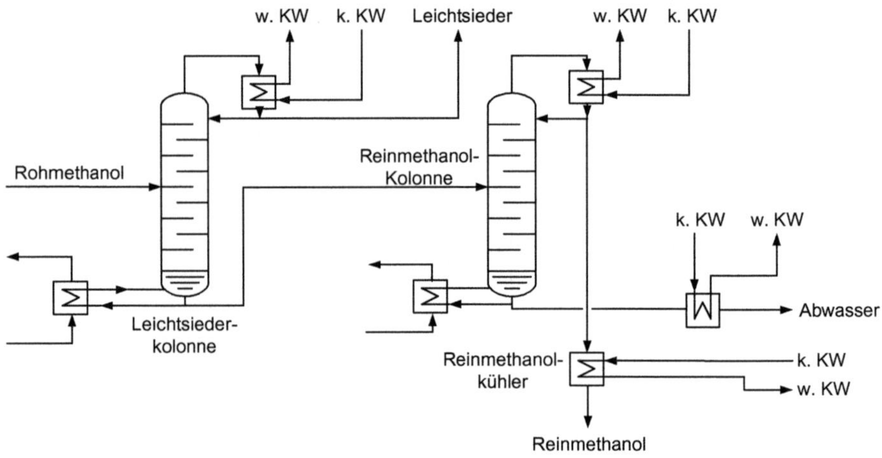

Рисунок 30 Схема очистки метанола с двумя ректификационными колонами

Таблица 21 Рамочные условия при моделировании процесса очистки метанола

показатель	ед. изм.	значение
Давление в исходном резервуаре	атмосфер	5
Давление в 1-ой ректификационной колоне	атмосфер	3
Число теоретических тарелок в 1-ой ректификационной колоне		20
Давление в о2-ой ректификационной колоне	атмосфер	1
Число теоретических тарелок во 2-ой ректификационной колоне		30
Давление пара для обогрева колон	атмосфер	5

глубокой очистки. Глубокая очистка может быть проведена в двух колонах, что однако требует более высоких капитальных вложений. Преимуществом является улучшенная энергоэффективность, удельная потребность в паре на 60% ниже, чем в случае использования только одной ректификационной колоны глубокой очистки. Как правило очистку метанола на трех колонах проводят на метанольных установках с большей мощностью. Так как мощность установки, рассмотренной в данном исследовании, относительно низкая, целесообразно использование только двух колон. Реализованная при моделировании схема очистки метанола изображена на Рисунке 30. Наиболее важные параметры моделирования собраны в Таблице 21.

6.4. **Производство синтетического бензина**

Для конверсии метанола в синтетический бензин предложено несколько технологий. В данном исследовании рассматривается технология получения бензина одностадийным процессом MtG, разработанным компанией Мобил.

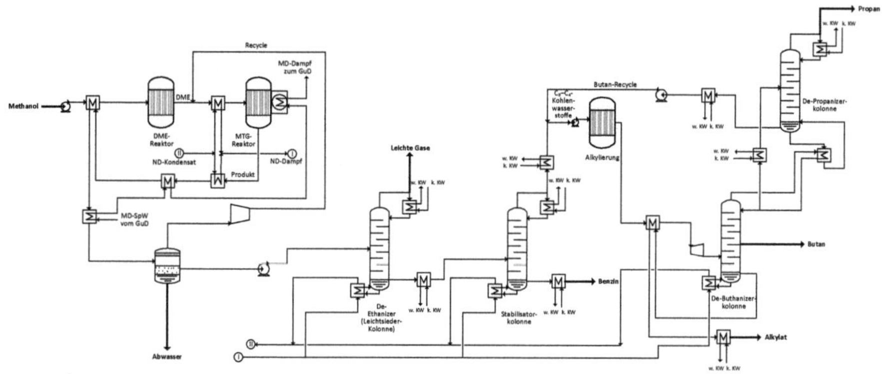

Рисунок 31 Схема процесса MtG

Таблица 22 Рамочные условия при моделировании процесса MtG

показатель	ед. изм.	значение
Температура на входе в реактор синтеза диметилового эфира	°C	300
Давление реактора синтеза диметилового эфира	атмосфер	22
Температура на входе в реактор синтеза бензина	°C	330
Давление реактора синтеза бензина	атмосфер	21
Температура на входе в процесс изомеризации полученного сырого бензина	°C	40
Давление процесса изомеризации бензина	атмосфер	14

По этой технологии метанол сначала подвергается дегидратации с образованием диметилового эфира. Для этого может использоваться сырой метанол, без предварительной очистки (90 вес.-% метанола). Поэтому отпадает необходимость в очистке метанола, описанной в Разделе 6.3, что улучшает экономические показатели технологической цепочки производства синтетического бензина. Конверсия метанола проводится в адиабатическом реакторе с фиксированным слоем катализатора Al_2O_3. Полученный диметиловый эфир без промежуточного охлаждения смеси подается в реактор синтеза бензина. Здесь также используется адиабатический реактор с фиксированным слоем катализатора (в данном случае цеолит-катализатор ZSM-5). Начальная температура на входе в реактор составляет 320–340 °C. Тепло реакции отводится из реактора через теплообменники, с генерацией пара высокого давления. Смесь на выходе из реактора разделяется после декомпрессии на три фазы (газ, водная фаза, органическая фаза). Органическая фаза разделяется с использованием четырех ректификационных колон на несколько фракций. Легкие газы и пропан-бутановая смесь отводится с головы колоны. Полученная бензиновая фракция подвергается

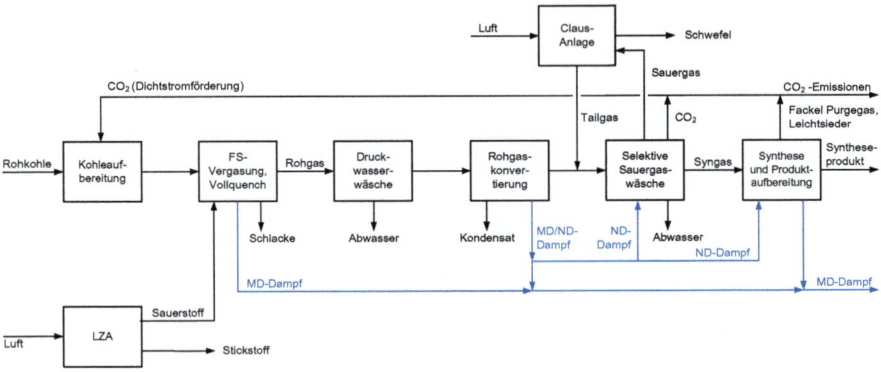

Рисунок 32 Блок-схема полной технологической цепочки на базе процесса газификации по технологии СИМЕНС

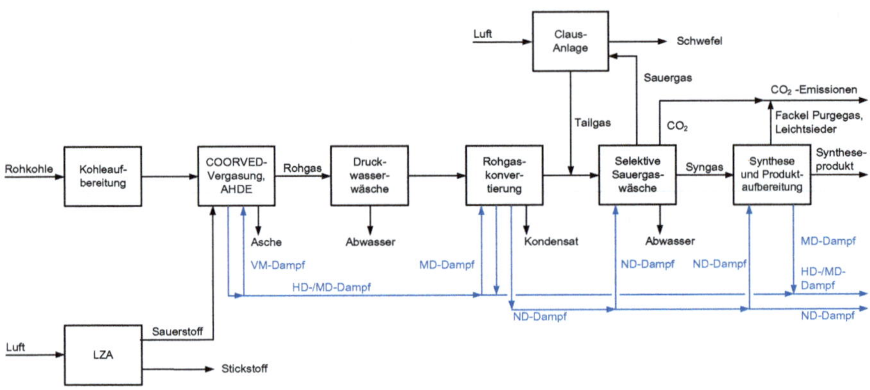

Рисунок 33 Блок-схема полной технологической цепочки на базе процесса газификации по технологии COORVED

дальнейшей обработке, включая процессы гидрирования и изомеризации с получением разветвленных насыщенных углеводородов. Реализованная при моделировании схема конверсии метанола в бензин, включая процессы обработки бензиновой фракции, изображена на Рисунке 31. Основные параметры, использованные при моделировании, представлены в Таблице 22.

6.5. Полные технологические цепочки

Для моделирования полных технологических цепочек использованы комбинации отдельных процессов, схемы и параметры моделирования которых описаны в Разделах 6.1 до 6.4. На Рисунке 32 изображена блок-схема цепочки, основанной на процессе газификации по технологии СИМЕНС. Синим цветом отмечены пути передачи пара между отдельными процессами и установками, целью которой является оптимизация теплового баланса цепочки. Соответствующая схема реализации технологической цепочки на базе технологии COORVED показана на Рисунке 33. Чтобы наглядно изобразить отмеченные в Разделе 6.1 преимущества технологии COORVED по сравнению с технологиями газификации в кипящем слое, связанные со снижением капитальных затрат ввиду отпавшей необходимости в некоторых узлах и элементах установок, на Рисунке 34 показана для сравнения блок-схема цепочки, основанной на газификации в кипящем слое.

7. Техническая и экономическая оценка

Для описанных в Разделе 6 технологических цепочек была проведена оценка их потенциала реализации, с учетом технических и экономических параметров. Обозначения всех восьми рассмотренных цепочек даны на Рисунке 24.

7.1. Техническая оценка

Определяющее значение для оценки применимости технологий глубокой переработки углей имеют показатели, описывающие процесс получения синтез-газа в газогенераторе. При этом выбор газификации определяется характеристиками перерабатываемого угля. В Таблице 23 приведены некоторые характеристики угля, параметры процесса газификации и рассчитанные специфические показатели для оценки выбранных технологий газификации, а также состав полученного синтез-газа для рассмотренных цепочек.

Результаты моделирования четко продемонстрировали многочисленные преимущества технологии COORVED для газификации рассмотренных углей. В большой степени это связано с относительно низкой температурой газификации, характерной для технологии COORVED. Благодаря этому технология COORVED отличается пониженными энергозатратами, необходимыми для поддержания температурного уровня газификации. Коммерческие технологии поточной газификации угля, в том числе и технология СИМЕНС, работают при температурах, лежащих около 150 °C выше температуры жидкого состояния, чтобы гарантировать полный переход золы в жидкое состояние. Несмотря на низкую температуру газификации по технологии COORVED, остаточное содержание углерода в золе, выходящей из газогенератора, удается понизить до уровня, не превышающего 1,2 вес.-%. Необходимо отметить, что в отдельных зонах реактора COORVED, прежде всего в зоне подвода первичного окислителя, наблюдаются температуры, превышающие температуру газа на выходе из реактора. Однако достигаемые температуры не превышают в значительной степени температуру жидкого состояния угля. Этим технология COORVED отличается от технологий поточной газификации (например технологии СИМЕНС). Потребность O_2 в случае технологии COORVED лежит на 18% (уголь ТА) и 22% (уголь КО)

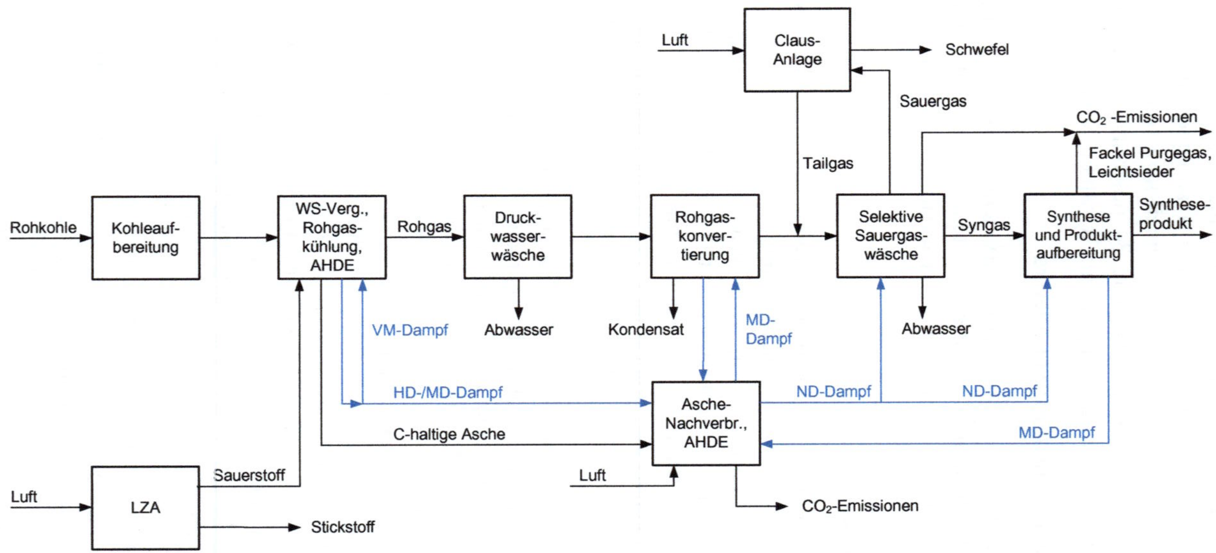

Рисунок 34 Блок-схема полной технологической цепочки на базе процесса газификации в кипящем слое

Таблица 23 Результаты моделирования газогенераторов

Уголь		ТА		КО	
Технология газификации		СИМ	COORVED	СИМ	COORVED
Показатель	Единицы измерения	Значение			
Температура размягчения	°C	1.440	1.440	1.310	1.310
Температура жидкого состояния	°C	1.500	1.500	1.410	1.410
Температура газа на выходе из газогенератора	°C	1.650	1.200	1.550	1.200
Давление газогенератора	атмосфер	40	40	40	40
Потребность O_2	$м^3$ (н.у.)/кг угля (сухое беззольное топливо)	0,711	0,583	0,774	0,606
Потребность пара	кг/кг угля (сухое беззольное топливо)	<0,01	0,361	<0,01	0,379
КПД холодного газа	%	77,7	84,9	75,8	84,2
Степень конверсии углерода	%	99,5	99,8	99,5	99,2
Выход синтез-газа	$м^3$ H_2+CO/кг угля (сухое беззольное топливо)	2,020	2,210	2,129	2,371
H_2/CO	моль/моль	0,315	0,649	0,362	0,747
Состав синтез-газа	На выходе из газогенератора				
H_2O	об.-% (сухой газ)	9,4	10,0	11,4	10,4
H_2	об.-% (сухой газ)	21,8	35,9	23,8	39,3
CO	об.-% (сухой газ)	69,3	55,4	65,8	52,6
CO_2	об.-% (сухой газ)	7,5	7,2	8,9	6,6
CH_4	об.-% (сухой газ)	0,0	0,2	0,0	0,3

ниже, чем по технологии СИМЕНС, ввиду более низкой температуры газификации. Данный эффект более значителен для высокозольных углей, т.к. отпадает потребность в подводе тепла для полного расплавления золы. Для примера, тепло необходимое для перевода золы в жидкое состояние (при оценочном значении энтальпии плавления золы 680 кДж/кг) составляет более одного процента теплотворной способности подведенного потока угля. Кроме того, технология COORVED характеризуется более высоким выходом синтез-газа, т.к. ввиду пониженной потребности в частичном окислении угля для генерации тепла, необходимого для достижения высоких температур газификации и перевода золы в жидкое состояние, снижается доля выхода CO_2. Помимо увеличенного значения выхода синтез-газа, технология COORVED отличается в положительную сторону по сравнению с технологией СИМЕНС и благодаря более высокому соотношению H_2/CO, превышающему в два раза соответствующее значения для поточной газификации. Это имеет большое значение прежде всего для переработки синтез-газа в химические продукты или синтетическое жидкое топливо, так как понижается необходимость в проведении реакции CO-шифт для получения заданного соотношения H_2/CO. В связи с этим снижается степень конверсии CO в CO_2 и как следствие возрастает количество метанола и бензина, произведенного на одну тонну угля, улучшается значение термического КПД процесса, увеличивается степень включения углерода в конечный товарный продукт, при одновременно пониженном значении выхода CO_2.

Как и следует ожидать, КПД холодного синтез-газа в случае высокозольных углей (уголь КО) ниже, чем в случае угля с более низким содержанием золы. Но необходимо отметить, что снижение КПД в случае технологии СИМЕНС было особенно значительным, несмотря на то, что температура газификации высокозольного угля КО по технологии СИМЕНС была задана в рамках моделирования на 100 °C ниже, чем для угля ТА с более низким содержанием золы. Интересно, что выход синтез-газа для низкокачественного угля КО выше, чем для угля ТА. Этот эффект был найден для обоих технологий газификации. Объяснением служит более высокое соотношение H/C и более низкое содержание кислорода в исходном угле (см. Таблицу 1).

Для проведения технической оценки пригодности углей для глубокой обработки рассматриваются три показателя, заданные Уравнениями 7–9. При расчете термического КПД по Уравнению (8) учитывается, что побочные продукты (отходы и отработанные газы) могут быть использованы для генерации пара (тепловая эффективность 90%). При пересчете теплотворной способности угля в электрическую энергию был принят КПД 38,5%.

КПД переработки угля: $$\eta_{\text{переработка}} = \frac{\dot{m}_{\text{MeOH}} \cdot Hu_{\text{MeOH}}}{\dot{m}_{\text{уголь,газификция}} \cdot Hu_{\text{уголь}}} \qquad (7)$$

Термический КПД: $$\eta_{\text{терм}} = \frac{\dot{m}_{\text{MeOH}} \cdot Hu_{\text{MeOH}} + 0{,}9 \cdot \dot{m}_{\text{побочн}} \cdot Hu_{\text{побочн}} + P_{\text{терм,избыт.пар}}}{\dot{m}_{\text{Куголь,газификация}} \cdot Hu_{\text{уголь}} + \left(\frac{P_{\text{электр,вспом}}}{\eta_{\text{еэлектр,нетто}}}\right)} \qquad (8)$$

Степень включения углерода в продукт: $$\eta_C = \frac{\dot{n}_{\text{C,MeOH}}}{\dot{n}_{\text{C,уголь,газификация}}} \qquad (9)$$

Результаты расчета материального и энергетического баланса рассмотренных технологических цепочек приведены в Таблице 24 для производства метанола и в Таблице 25 для производства бензина. Все расчеты проведены для установки с термической мощностью 500 МВт (соответствует теплотворной способности подводимого угля). Видно, что технология COORVED имеет значительные преимущества по сравнению с технологией СИМЕНС.

Потребность энергии для размельчения угля в случае поточной технологии, как например СИМЕНС, выше чем для технологии COORVED. Кроме того,

Таблица 24 Материальный и энергетический баланс для технологических цепочек, включающих синтез метанола

Уголь		ТА		КО	
Технология газификации		СИМ	COORVED	СИМ	COORVED
Показатель	Единицы измерения	Значение			
Подвод угля	МВт(терм) кг/сек тонн/день тысяч тонн/год	500 20,9 1.811 661,0	500 20,9 1.811 661,0	500 25,4 2.192 800,1	500 25,4 2.192 800,1
Выход метанола	МВт(терм) кг/сек тонн/день тысяч тонн/год	280,6 14,1 1.217 444,7	314,2 15,8 1.363 498,3	274,0 13,8 1.189 435,2	312,8 15,7 1.357 495,1
Удельный выход продукта	кг метанола/кг угля	0,675	0,756	0,543	0,618
Потребность в электроэнергии на:					
- подготовку угля	МВт(электр)	1,13	0,57	1,37	0,69
- разделение воздуха	МВт(электр)	20,69	17,11	20,83	16,46
- газификацию	МВт(электр)	4,43	2,58	4,43	2,58
- очистку газа	МВт(электр)	4,93	4,58	4,76	4,28
- синтез метанола	МВт(электр)	3,63	4,01	3,70	3,97
всего	МВт(электр)	34,81	28,83	35,09	27,98
Удельная потребность в электроэнергии	МДж/кг метанола	2,47	1,83	2,55	1,78
Теплотворная способность отходов	МВт(терм)	23,78	25,76	21,70	24,19
Генерация пара					
- высокого давления	МВт(терм)	0,00	42,61	0,00	43,45
- среднего/низкого давл.	МВт(терм)	112,30	33,28	118,43	33,76
- из отходов	МВт(терм)	21,40	23,19	19,53	21,77
сумма	МВт(терм)	133,70	99,07	137,96	98,98
Выход CO_2:					
- при промывке газа	кг/сек	25,08	23,34	22,73	20,43
- при рециркуляции	кг/сек	1,13	1,07	1,02	0,96
- в ректификац. колоне	кг/сек	1,41	1,44	1,37	1,38
всего	кг/сек	27,62	25,85	25,11	22,77
Выход CO_2 на кг продукта	кг/кг метанола	1,96	1,64	1,83	1,45
КПД переработки угля	%	56,1	62,8	54,8	62,6
Термический КПД	%	69,1	71,0	68,6	71,0
Степень включения углерода в продукт	%	40,4	45,3	42,0	48,0

Таблица 25 Материальный и энергетический баланс для технологических цепочек, включающих производство синтетического бензина

Уголь		TA		KO	
Технология газификации		СИМ	COORVED	СИМ	COORVED
Показатель	Единицы измерения	Значение			
Подвод угля	МВт(терм) кг/сек тонн/день тысяч тонн/год	500 20,9 1.811 661,0	500 20,9 1.811 661,0	500 25,4 2.192 800,1	500 25,4 2.192 800,1
Выход бензина	МВт(терм) кг/сек тонн/день тысяч тонн/год	255,3 4,9 423 154,5	289,8 5,6 484 176,6	249,7 4,8 423 151,4	289,8 5,6 484 176,6
Удельный выход продукта	кг бензина/кг метанола кг бензина/кг угля	0,310 0,259	0,310 0,294	0,310 0,208	0,310 0,242
Побочные продукты - высшие спирты - пропан - бутан	 кг/сек кг/сек кг/сек	 0,17 0,30 0,51	 0,20 0,34 0,58	 0,17 0,29 0,50	 0,20 0,34 0,58
Потребность в электроэнергии на: - подготовку угля - разделение воздуха - газификацию - очистку газа - синтез метанола - синтез бензина всего	 МВт(электр) МВт(электр) МВт(электр) МВт(электр) МВт(электр) МВт(электр) МВт(электр)	 1,13 20,69 4,43 4,93 3,63 1,22 36,03	 0,57 17,11 2,58 4,58 4,01 1,23 30,06	 1,37 20,83 4,43 4,76 3,70 1,22 36,31	 0,69 16,46 2,58 4,28 3,97 1,23 29,20
Удельная потребность в электроэнергии	МДж/кг бензина	6,13	4,50	6,31	4,37
Теплотворная способность отходов	МВт(терм)	23,78	25,76	21,70	24,19
Генерация пара - высокого давления - среднего/низкого давл. - из отходов сумма	 МВт(терм) МВт(терм) МВт(терм) МВт(терм)	 0,00 106,06 28,39 134,99	 42,61 67,92 28,37 138,9	 0,00 110,86 26,19 137,05	 43,45 68,72 26,31 138,48
Выход CO_2: - при промывке газа - при рециркуляции - при синтезе бензина всего	 кг/сек кг/сек кг/сек кг/сек	 25,08 1,13 1,14 27,35	 23,34 1,07 1,29 25,70	 22,73 1,02 1,11 24,86	 20,43 0,96 1,29 22,68

(continued)

Таблица 25 (continued)

Уголь		ТА		КО	
Технология газификации		СИМ	COORVED	СИМ	COORVED
Показатель	Единицы измерения	Значение			
Выход CO_2 на кг продукта	кг/кг бензина	4,65	3,85	4,32	3,40
КПД переработки угля	%	51,0	58,0	49,9	58,0
Термический КПД	%	64,3	65,4	63,6	65,5
Степень включения углерода в продукт	%	38,4	43,6	40,0	46,4

по технологии СИМЕНС требуется дополнительная электроэнергия для увеличения давления CO_2, необходимого для подвода угольной пыли в газогенератор, от 1 до 40 атмосфер. В общей сумме, потребность в электроэнергии для технологии СИМЕНС на 17–20% выше, чем для технологии COORVED. В расчете на кг продукта, различие между обоими технологиями еще выше и лежит в диапазоне 26–30%. Это связано со значительно более высоким выходом метанола в рамках технологии COORVED.

Термический КПД, так же как и КПД переработки угля в продукт и степень включения углерода в конечный продукт, для технологических цепочек, включающих процесс COORVED, превышает соответствующие значения рассчитанные для технологических цепочек, основанных на использовании технологии газификации СИМЕНС. Расчеты в случае COORVED приводят к более высоким показателя производства пара, благодаря использованию конвекционного охладителя синтез-газа. Правда необходимо учитывать, что, в отличие от технологии СИМЕНС, часть пара идет на проведение реакции CO-шифт. Тем не менее, термический КПД полной технологической цепочки в случае COORVED достигает более высокие значения.

7.2. **Экономическая оценка**

На базе результатов моделирования полных технологических цепочек (расчета материального и энергетического баланса) была проведена экономическая оценка цепочек. Расчет капитальных вложений был проведен по методу масштабирования мощностей. Себестоимость продукта рассчитана в среднем на весь период амортизации. Рамочные условия и допущения, сделанные при расчете экономических показателей, собраны в Таблице 26 и Таблице 27.

В Таблице 28 представлены заложенные при расчетах данные по капитальным вложениям для базовых установок, с указанием вида сырья/продукта, на основе которого производилось масштабирование.

Таблица 26 Рамочные условия экономической оценки [79,80]

Показатель	Значение
Общие рамочные условия	
Базовый год	2015
Начало работ по проектированию	01.01.2016
Начало работ по строительству	01.01.2017
Ввод мощностей в эксплуатацию	01.01.2020
Период амортизации	20 лет
Индекс инфляции	2 % в год
Рост стоимости	1 % в год
Структура финансирования и кредитования	
Собственный капитал акционеров	33 %
Процент на капитал акционеров	17,5 %
Заемный капитал	67 %
Процент на заемный капитал	6,7 %
Средний процент	10 % в год
Распределение капитальных вложений во время строительства	
2017-2018	10 %
2018-2019	60 %
2019-2020	30 %
Фиксированные эксплуатационные расходы (в процентах по отношению к сумме инвестиции)	
Страхование	0,5 %
Техническое обслуживание и ремонт	5 %
Расходы на содержание персонала	86.208 евро в год

Таблица 27 Допущения относительно затрат и доходов [79, 80, 81]

Показатель	Год	Ед. изм.	Значение
Переменные эксплуатационные расходы			
Рядовой уголь, шахта «Комсомолец»	2015	евро/т	20,00
Рядовой уголь, шахта «Талдинская-Западная-2»	2015	евро/т	25,00
Электроэнергия	2015	евро/МВт час	42,00
Техническая вода	2008	евро/т	0,20
Затраты на утилизацию отходов			
Сточные воды	2015	евро/т	0,74
Зола/шлак	2008	евро/т	0,00
Штрафы на выброс CO_2	2015	евро/т	10,00
Доходы от сбыта побочных продуктов			
Сера	2010	евро/т	149,60
Пропан/Бутан	2015	евро/т	553,40

Таблица 28 База данных для оценки капитальных вложений по методу масштабирования мощностей [79, 80, 82]

Процессы	База для масштабирования	Затраты милл. евро	Мощность кг/сек
Поточный газогенератор СИМЕНС	Подача угля	192	40,0
Газогенератор HTW	Подача угля	248	46,0
Газогенератор COORVED	Подача угля	208	46,0
Установка по разделению воздуха	Полученный кислород	180	95,4
Установка по очистке газа	Выделенный CO_2	101	181,9
Синтез метанола	Полученный метанол	6	35,6
Очистка метанола	Подача сырого метанола	3	11,2
Производство бензина	Подача сырого метанола	4	10,6

Базой для оценки капитальных вложений для газогенератора COORVED служит коммерческий газогенератор HTW. Благодаря более компактной конструкции газогенератора COORVED, капитальные вложения оцениваются на 30% ниже по сравнению с HTW. Так как технология COORVED пока продемонстрирована только на опытной установке, для нее учтен фактор технологического риска 1,2, как это принято в подобных случаях.

Полученные значения для необходимых капитальных вложений представлены в Таблице 29 для цепочек производства метанола и в Таблице 30 для производства синтетического бензина. Указан также вклад различных составляющих в общую сумму капитальных вложений.

Таблица 29 Общие капитальные вложения (в миллионах евро) для технологических цепочек производства метанола

Цепочка	ТА-СИМ-MeOH	ТА-COORVED-MeOH	КО-СИМ-MeOH	КО-COORVED-MeOH
Подготовка угля	68,114	65,264	77,843	74,587
Разделение воздуха	59,848	52,383	60,138	51,000
Газогенератор	128,969	126,381	147,390	144,435
Очистка газа	62,150	51,318	63,453	50,299
Синтез и очистка продукта	9,183	9,705	9,040	9,638
Инфраструктура и вспомогательные установки	44,018	22,482	50,305	25,693
Производственный комплекс	372,282	327,532	408,170	355,651
Общая сумма	542,871	478,724	594,477	519,627

Таблица 30 Общие капитальные вложения (в миллионах евро) для технологических цепочек производства синтетического бензина

Цепочка	TA-СИМ-MtG	TA-COORVED-MtG	KO-СИМ-MtG	KO-COORVED-MtG
Подготовка угля	68,114	65,264	77,843	74,587
Разделение воздуха	59,848	52,383	60,138	51,000
Газогенератор	128,969	126,381	147,390	144,435
Очистка газа	62,150	51,318	63,452	50,299
Синтез и очистка продукта	35,409	38,000	34,894	37,958
Инфраструктура и вспомогательные установки	44,018	22,482	50,305	25,693
Производственный комплекс	398,507	355,827	434,024	383,971
Общая сумма	580,321	518,733	631,347	558,555

Сравнение между обоими углями показывает, что необходимые капитальные вложения для технологических цепочек переработки высокозольного угля КО являются более высокими, чем для угля ТА. В отношении технологий газификации расчеты показывают, что капитальные вложения для цепочек, основанных на использовании технологии поточной газификации СИМЕНС, выше соответствующих значений для технологии COORVED приблизительно на 5–7%. Вклады составляющих звеньев в общую сумму инвестиций можно лучше понять и обосновать с учетом технического анализа технологических цепочек. Технология СИМЕНС требует более сильного измельчения угля, ввиду чего соответствующие капитальные вложения превышают значения для технологии COORVED примерно на 4%. Расходы на установку по разделению воздуха в случае технологии COORVED ниже на 12–15% из-за меньшей удельной потребности в кислороде. Даже с учетом фактора технологического риска, капитальные вложения для газогенератора COORVED ниже чем для поточного газогенератора СИМЕНС. Кроме того технология COORVED отличается меньшими капитальными затратами в системы вывода шлаков и дожига остаточного углерода в шлаке (эти затраты учтены в строке «инфраструктура и вспомогательные установки») и менее дорогостоящими установками по очистке синтез-газа. Капитальные затраты на установки по синтезу метанола, очистке метанола и переработке метанола в бензин в случае технологических цепочек, основанных на технологии COORVED, несколько выше, чем для технологии СИМЕНС. Но это связано с тем, что технология COORVED отличается более высоким удельным выходом синтез-газа и соответственно более высоким удельным выходом метанола и бензина.

Исходя из капитальных вложений, была проведена оценка себестоимости полученного продукта. Рассчитанные значения представлены в Таблице 31

Таблица 31 Оценка себестоимости продукта (в евро) для технологических цепочек производства метанола

Цепочка	ТА-СИМ-MeOH	TA-COORVED-MeOH	КО-СИМ-MeOH	KO-COORVED-MeOH
Капитальные вложения	34,30	27,01	38,47	29,44
Стоимость угля	7,70	6,88	7,64	6,69
Эксплуатационные расходы, включая обслуживание	19,99	16,47	22,27	18,26
Необходимый годовой доход	62,00	50,36	68,38	54,39
Учет дохода от сбыта побочных продуктов	-5,77	-3,97	-6,07	-3,99
Себестоимость продукта - в евро/МВт·час продукта - в евро/тонну продукта - в евро/литр продукта	 56,24 310,86 0,25	 46,39 256,45 0,20	 62,31 344,44 0,27	 50,40 278,62 0,22

Таблица 32 Оценка себестоимости продукта (в евро) для технологических цепочек производства синтетического бензина

Цепочка	ТА-СИМ-MtG	TA-COORVED-MtG	КО-СИМ-MtG	KO-COORVED-MtG
Капитальные вложения	40,31	31,73	44,84	34,17
Стоимость угля	8,47	7,46	8,38	7,22
Эксплуатационные расходы, включая обслуживание	22,66	17,71	25,01	17,79
Необходимый годовой доход	71,44	56,90	78,23	59,18
Учет дохода от сбыта побочных продуктов	-13,15	-10,91	-13,38	-10,86
Себестоимость продукта - в евро/МВт·час продукта - в евро/тонну продукта - в евро/литр продукта	 58,29 697,17 0,54	 45,99 550,01 0,43	 64,85 775,63 0,60	 48,32 577,93 0,45

для цепочек производства метанола и в Таблице 32 для производства синтетического бензина. Указан также вклад различных составляющих. При расчете себестоимости продукта был учтен ожидаемый доход от сбыта побочных продуктов.

Как и следует ожидать, себестоимость продуктов глубокой переработки для низкозольного угля ТА ниже, чем для высокозольного угля КО. При расчетах была заложена на 5 евро/тонну более высокая стоимость для угля ТА по сравнению с углем КО. Тем не менее, более высокий удельный выход продукта приводит к тому, что вклад стоимости угля в себестоимость продукта ниже в случае переработки угля ТА. Кроме того, более низкие затраты на очистку синтез-газа приводят к тому, что и эксплуатационные расходы в случае угля ТА ниже, чем для угля КО.

Сопоставление технологий газификации показывает, что себестоимость продуктов, основанных на использовании технологии COORVED ниже чем в случае газификации СИМЕНС. Вклады составляющих звеньев в общую сумму инвестиций можно лучше понять и обосновать с учетом технического анализа технологических цепочек. Определяющими факторами являются более низкие капитальные вложения, более высокий удельный выход продуктов (и соответственно более низкие вклад стоимости угля) и меньшие затраты на очистку полученного синтез-газа (более низкие эксплуатационные расходы). Технологические цепочки на основе газификации СИМЕНС правда отличаются более высоким доходом от сбыта побочных продуктов, ввиду повышенного выхода пара высокого и среднего давления. Однако этот вклад в общую себестоимость продукта относительно невелик.

Что качается технологических цепочек производства синтетического бензина, можно отметить, что здесь может быть достигнут существенно более высокий доход от сбыта побочных продуктов, чем в случае производства метанола. Однако капитальные вложения и эксплуатационные расходы также выше, ввиду необходимости проведения дополнительного процесса синтеза.

Методом анализа чувствительности была проведена оценка влияния различных параметров на себестоимость продукта. Для примера в Таблице 33 представлены рассчитанные значения себестоимости синтетического бензина

Таблица 33 Результаты анализа чувствительности для цепочки TA-COORVED-MtG

Показатель	Ед. изм.	Нижняя граница	Базовое значение	Верхняя граница
Капитальные вложения	милл. евро	189,9 (-50 %)	379,9	569,9 (+50 %)
Себестоимость продукта	евро/литр	0,27	0,43	0,58
Годовой ресурс работы	%	80 (-10 %)	88,7	95 (+7 %)
Себестоимость продукта	евро/литр	0,47	0,43	0,41
Доход от сбыта пропан/бутана	евро/тонну	276,7 (-50 %)	553,4	830,1 (+50 %)
Себестоимость продукта	евро/литр	0,46	0,43	0,40
Стоимость электроэнергии	евро/МВт·час	15 (-64 %)	42	100 (+138 %)
Себестоимость продукта	евро/литр	0,42	0,43	0,44

для цепочки TA-COORVED-MtG, полученные при вариации определенных параметров в заданном диапазоне. Видно, что объем капитальных вложений оказывает определяющее влияние на себестоимость продукта. Важное значение имеет также годовой ресурс работы оборудования. Доход от сбыта побочной продукции играет весомую роль только при очень высоком росте цен. Стоимость электроэнергии почти не оказывает влияние на себестоимость произведенного бензина.

8. Заключение

В данном исследовании на основе анализа различных технологий газификации угля были выбраны две технологии, наиболее пригодные для глубокой переработки российских углей различного качества (низкозольные угли, высокозольные угли) – поточная газификация СИМЕНС и новая технология COORVED, специально разработанная для низкокачественных углей. Кроме того, в исследовании был дан обзор, с учетом технических и экономических аспектов, различных путей переработки синтез-газа в конечную продукцию и выбраны два продукта с высокой добавленной стоимостью – метанол и синтетический бензин высокого качества. При помощи коммерческого программного пакета Aspen Plus® проведена оптимизация полных технологических цепочек глубокой переработки для двух выбранных углей – низкозольного угля (шахта «Талдинская-Западная-2») и высокозольного угля (шахта «Комсомолец»). Рассчитанные технические и экономические показатели позволяют провести обоснованную оценку возможностей внедрения глубокой переработки угля в российской промышленности, принимая во внимание экономическую ситуацию в угольных регионах России (например в Кузбассе) и с учетом конкретных характеристик перерабатываемого угля.

Расчеты материального баланса для приводят к следующим результатам:

– для низкозольного угля шахты «Талдинская-Западная-2»

технология СИМЕНС:
1 тонна рядового угля -> 675 кг метанола -> 259 кг бензина
технология COORVED:
1 тонна рядового угля -> 756 кг метанола -> 294 кг бензина

– для высокозольного угля шахты «Комсомолец»

технология СИМЕНС:
1 тонна рядового угля -> 543 кг метанола -> 208 кг бензина
технология COORVED:
1 тонна рядового угля -> 618 кг метанола -> 242 кг бензина

В обоих случаях (низкозольный уголь, высокозольный уголь) технология COORVED имеет значительные преимущества по сравнению с коммерческой технологией поточной газификации, обусловленные инновационным характером процесса COORVED.

К аналогичным выводам приводит сопоставление рассчитанной себестоимости продукта:

– для низкозольного угля шахты «Талдинская-Западная-2»

технология СИМЕНС:
311 евро/тонну метанола 0,54 евро/литр бензина
технология COORVED:
256 евро/тонну метанола 0,43 евро/литр бензина

– для высокозольного угля шахты «Комсомолец»

технология СИМЕНС:
344 евро/тонну метанола 0,60 евро/литр бензина
технология COORVED:
279 евро/тонну метанола 0,45 евро/литр бензина

Проведенный анализ позволяет сделать вывод, что глубокая переработка угля с получением метанола имеет определенный потенциал реализации в условиях российской промышленности. Экономические расчеты технологических цепочек на базе инновационной технологии газификации COORVED приводят к значениям себестоимости метанола в районе 250 евро/тонну для низкозольного угля и 280 евро/тонну для высокозольного угля. Цена метанола на мировых рынках находилась в 2010–2015 годах в районе 300–500 евро/тонну. Лишь в 2016 году она опустилась ниже 250 евро/тонну в связи с обострением мирового экономического кризиса. На начало 2017 года опять наблюдается рост цен на метанол, в Европе цена составляет 370 евро/тонну, в Китае 430 евро/тонну. С учетом географического положения основных угольных регионов России и возможностей выхода продукции на азиатские рынки интересно отметить, что цены на метанол в Азии в последние 15 лет стабильно лежат выше цен в Европе, разница в цене временами существенно превышает 100 евро/тонну.

Необходимо отметить, что проведенные расчеты основаны на ряде оптимистичных допущений. Например не учтено, что в крупномасштабном производстве обычно требуются дополнительные (резервные) газогенераторы, что приводит к росту капитальных вложений. Можно ожидать, что реальная себестоимость продукта глубокой переработки угля будет превышать расчеты примерно на 10–20%. В этой связи особо важно учитывать возможный потенциал снижения затрат, связанный с развитием кластеров, объединяющих углехимию и угольную энергетику. Здесь в свою очередь заложен потенциал снижения себестоимости продукта на 10–15%. Этот аспект также рассмотрен в данном исследовании.

В целом можно сделать заключение, что при условии применения передовых технологий газификации угля имеется определенный потенциал для глубокой переработки углей, в том числе и высокозольных углей, в условиях российской промышлености. Однако предварительная оценка применимости различных цепочек переработки угля показала, что рассчитанные значения себестоимости продукта (метанола) лежат на границе самоокупаемости. В этой связи представляется целесообразным проведение более углубленного анализа экономической целесообразности различных вариантов комплексного подхода к углехимии, с привлечением детальной информации по рамочным условиям для выбранного региона и намеченных рынков сбыта.

9. Приложение

9.1. Список рисунков

9.2. **Список таблиц**

10. Литература

1. Данные по техническим характеристикам российских углей предостав лены Санкт-Петербургским Горным университетом.
2. BP Statistical review of world energy 2015. http://www.bp.com/content/dam/bp/pdf/energy-economics/statistical-review-2015/bp-statistical-review-of-world-energy-2015-full-report.pdf (на 11.01.2017).
3. Добыча топливно-энергетических полезных ископаемых. ГосКомСтат. http://www.gks.ru/free_doc/new_site/business/prom/natura/natura2g.htm (на 11.01.2017).
4. Mochalnikov SV. Current status and development prospects of coal industry in Russia. International Symposium "Clean Coal Day", Tokyo, 08.-10.09.2015. http://www.jcoal.or.jp/coaldb/shiryo/material/upload/1-12speech%203%20Russia%20Mr.%20Mochalnikov.pdf (на 11.01.2017).
5. Higman C, Burgt M. Gasification. Amsterdam, Gulf Professional; 2003.
6. Gräbner M. Industrial coal gasification technologies covering baseline and high-ash coal. Weinheim, Wiley-VCH; 2014.
7. Higman C. State of the gasification industry: worldwide gasification database 2015 update. Colorado Springs; 2015.
8. Timmermann H. Mk Plus—the next Generation Lurgi FBDB Gasification. Leipzig, Germany; 2012.
9. Olschar M, Schulze O. The BGL commercial plants and pilot testing. Leipzig, Germany; 2012.
10. Adlhoch W, Keller J, Herbert PK. The development of the HTW coal gasification process; 1990.
11. Xu J. ECUST coal gasification Update: new projects and operations experience. Washington, D.C; 2014
12. Zhuang Q, Bowers K. Unlocking potential value from low rank coal with TRIG gasification technology. Beijing; 2013.
13. Gräbner M, Meyer B. Performance and exergy analysis of the current developments in coal gasification technology. Fuel. 2014; 116:910–920.
14. Laugwitz A, Meyer B. New frontiers and challenges in gasification technologies. In: Nikrityuk PA, Meyer B, editors. Gasification processes: modeling and simulation. Hoboken, Wiley; 2014.
15. Schreiner M. Research guidance studies to assess gasoline from coal by methanol-to-gasoline and Sasol-type Fischer-Tropsch technologies: final

report. Mobil R&D Corp. U.S. DoE Report FE-2447-13. Department of Energy, August 1978, 460 p.

16. Methanex Investor Presentation, December 2016. https://www.methanex.com/sites/default/files/MEOH%20Presentation%20Dec%202016.pdf (на 11.01.2017).
17. Ott J, Gronemann V, Pontzen F, Fiedler E, Grossmann G, Kersebohm B, Weiss G, Witte C. Methanol. Ullmann's encyclopedia of technical chemistry, 7th ed. Wiley-VCH Verlag GmbH & C. KGaA, Weinheim; 2013.
18. Higman C, Tam S. Advances in Coal Gasification, Hydrogenation, and Gas Treating for the Production of Chemicals and Fuels. Chem. Rev. 2014; 114:1673–1708.
19. Mastrup M. XtL – the Topsoe approach. 4th International Freiberg Conference on IGCC & XtL Technologies, Dresden, 3.-6.05.2010. http://tu-freiberg.de/sites/default/files/media/professur-fuer-energieverfahrenstechnik-und-thermische-rueckstandsbehandlung-16460/publikationen/2010-01-4.pdf (на 11.01.2017).
20. Minchener AJ. Coal-to-oil, gas and chemicals in China. IEA Clean Coal Center. Report CCC/181. 2011. https://www.usea.org/sites/default/files/022011_Coal-to-oil,%20gas%20and%20chemicals%20in%20China_ccc181.pdf (на 11.01.2017).
21. Methanol and derivatives. Firmenschrift. https://www.engineering-airliquide.com/sites/activity_eandc/files/2016/07/13/methanol_and_derivatives_brochure-june_2016.pdf (Stand 11.01.2017).
22. Successful startup of Hainan Island methanol plant. Weblink. http://www.hydrocarbonprocessing.com/news/2011/02/successful-startup-of-hainan-island-methanol-plant-china (на 11.01.2017).
23. Tweddle G. Coal & natural gas chemicals. Tecnon OrbiChem Marketing Seminar. Asia Petrochemical Industry Conference, Pattaya, 15.05.2014 http://www.orbichem.com/userfiles/APIC%202014/APIC2014_02_Gillian_Tweddle.pdf (на 11.01.2017).
24. Methanex monthly average regional posted contract price history. https://www.methanex.com/sites/default/files/methanol-price/MxAvgPrice_Dec%2028%2C%202016.pdf (на 11.01.2017).
25. Johnson D. Global methanol market review. http://www.ptq.pemex.com/productosyservicios/eventosdescargas/Documents/Foro%20PEMEX%20Petroqu%C3%ADmica/2012/PEMEX_DJohnson.pdf (на 11.01.2017).
26. Methanol synthesis, the synthesis loop. SUPERMETHANOL Projekt im Rahmen des 7th European Commision framework programme. http://www.supermethanol.eu/index.php?id=22&rid=13&r=methanol_synthesis (на 11.01.2017).
27. Chen L, Jiang Q, Song Z, Posarac D. Optimization of methanol yield from Lurgi reactor. Chem. Engineering Techn. 2011; 34:817–822.
28. Haid J, Koss U. Lurgi's MegaMethanol technology opens the door for a new era in down-stream applications. Studies Surf. Sci. Catal. 2001; 136:399–404.

29. Wurzel T. Lurgi MegaMethanol technology – delivering the building blocks for future fuel and monomer demand. DGMK Conference "Synthesis Gas Chemistry", Dresden, 04.-06.10.2006 http://www.dgmk.de/petrochemistry/abstracts_content14/Wurzel.pdf (на 11.01.2017).
30. Heydorn EC, Diamond BW, Lilly RD. Commercial scale demonstration of the liquid phase methanol (LPMeOH) process. Final report DoE/NETL: DE-FC22-92PC90543. June 2003. https://www.netl.doe.gov/File%20Library/Research/Coal/major%20demonstrations/cctdp/Round3/LPMEOH/LPMEOH Final.pdf (на 11.01.2017).
31. Air Products and Chemicals Inc. Economic Analysis: LPMeOH process as an Add-on Integrated Gasification Combined Cycle for coproduction. Topical report DoE/NETL: DE-FC22-92PC90543. October 1998. https://www.netl.doe.gov/File%20Library/Research/Coal/major%20demonstrations/cctdp/Round3/LPMEOH/Econfin.pdf (на 11.01.2017).
32. Chang CL, Chu CTW, Socha RF. Methanol conversion to olefins over ZSM-5. I. Effect of temperature and zeolite SiO_2/Al_2O_3. J. Catal. 1984; 86:289–296.
33. Chang CL, Silvestri AJ. MTG. Origin, evolution, operation. CHEMTECH. 1987; 17:624–631.
34. Keil FJ. Methanol-to-hydrocarbon: process technology. Micropor. Mesopor. Mater. 1999; 29:49–66.
35. Topp-Jorgensen J, Topsoe integrated gasoline synthesis – the TIGAS process. Studies Surf. Sci. Catal. 1988; 36:293–305.
36. Kuschel M, Engelmann J, Meyer B. STF technology – production of high octane gasoline from synthesis gas on a new route. International Pittsburgh Coal Conference, Beijing; 2013 Sept 16–18.
37. Baitalow F, Stahlschmidt R, Seifert P, Pardemann R, Meyer B, Engelmann J. Production of gasoline from coal based on the STF process. 18th International Coal Preparation Congress, St. Petersburg; 2016 June 28–Jul 1.
38. Bertau M. Offermanns H, Plass L, Schmidt F, Wernicke HJ (ed.). Methanol: the basic chemical and energy feedstock of the future. Springer, Heidelberg; 2014.
39. China's Yulin Energy to start up new MTO unit in June 2014. Weblink. www.icis.com/resources/news/2013/10/10/9713958/china-s-yulin-energy-to-start-up-new-mto-unit-in-june-2014/ (на 11.01.2017).
40. KBR deploys its methanol-to-olefin recovery technology in China. Weblink. www.2b1stconsulting.com/kbr-deploys-its-methanol-to-olefin-recovery-technology-in-china/ (на 11.01.2017).
41. Chemiewirtschaft in Zahlen. VCI. August 2016. https://www.vci.de/vci/downloads-vci/publikation/chemiewirtschaft-in-zahlen-print.pdf (на 11.01.2017).
42. Chang CL. Hydrocarbons from methanol. Catal. Rev. Sci. Eng. 1983:1–118.
43. Grimmer HR, Thiagarajan N, Nitschke E. Conversion of methanol to liquid fuels by the fluid bed Mobil process (a commercial concept). Studies Surf. Sci. Catal. 1988; 36:273–291.

44. Gould RM, Tabak SA. Multi-stage system for converting oxygenates to liquid hydrocarbons with aliphatic recycle. U.S. Patent 4,689,205 A. 1987.
45. Tabak SA, Avidan AA, Krambeck FJ. Production of synthetic gasoline and diesel fuel from non-petroleum resources. ACS National Meeting. New York; 1986 Apr 13–18. https://web.anl.gov/PCS/acsfuel/preprint%20archive/Files/31_2_NEW%20YORK_04-86_0293.pdf (на 11.01.2017).
46. Liebner W, Wagner M. MtSynfuels, efficient and economical alternative to Fischer-Tropsch fuels. Erdöl, Erdgas, Kohle. 2004; 120:323–326.
47. Rothaemel M, Holtmann H.-D. Methanol to propylene MTP—Lurgi's way. Oil Gas European Magazine. 2002:27–30.
48. Koempel H, Liebner W. Lurgi's Methanol to propylene (MTP): report on a successful commercialisation. Studies Surf. Sci. Catal. 2007; 167:261–267.
49. Wurzel T, Koempel H, Rothaemel M, Liebner W. MTP—the on-purpose propylene technology for a changing feedstock environment. 19th World Petroleum Congress. Madrid; 2008 June 29–Jul 03.
50. Vora BV, Marker TL, Barger PT, Nilsen HR, Kvisle S, Fuglerud T. Economic route for natural gas conversion to ethylene and propylene. Studies Surf. Sci. Catal. 1988; 36:87–98.
51. Chen JQ, Bozzano A, Glover B, Fuglerud T, Kvisle S. Recent advancements in ethylene and propylene production using the UOP/Hyro MTO process. Catal. Today. 2005; 106:103–107.
52. Quann RJ, Green LA, Tabak SA, Krambeck FJ. Chemistry of olefin oligomerization over ZSM-5 catalyst. Ind. Eng. Chem. Res. 1988; 27:565–570.
53. Tabak SA, Krambeck FJ. Shaping process making fuels. Hydrocarbon Processing. 1985; 64:72–74.
54. Tabak SA. Production of synthetic diesel fuel from light olefins. AIChE National Meeting. Philadelphia; 1984 Aug 26–31.
55. Koempel H, Liebner W, Wagner M. Lurgi's gas to chemicals (GTC): advanced technologies for natural gas monetisation. Gastech. Bilbao; 2005 Mar 14–17. http://www.ivt.ntnu.no/ept/fag/tep4215/innhold/LNG%20Conferences/2005/SDS_TIF/050140.pdf (на 11.01.2017).
56. Koehler E, Schmidt F, Wernicke HJ, de Pontes M, Roberts HL. Converting olefins to diesel—the COD process. Hydrocarbon Techn. Int. Summer. 1995: 37–40.
57. Knottenbelt CD, Dunlop C, Zono K. Process for catalytic conversion of Fischer-Tropsch derived olefins into distillates. U.S. Patent 8,318,003 B2. 2012.
58. Knottenbelt CD, Ntshabele JPP. PetroSA's conversion of olefines to distillate (COD) research programm. 18th World Petroleum Congress. Johannesburg; 2005 Sept 25–29.
59. Steynberg A, Dry M (Ed.). Fischer-Tropsch technology. Chapter 5: Commercial Fischer-Tropsch process application. Studies Surf. Sci. Catal. 2004; 152:406–481.

60. SASOL Facts 2012/13. Брошюра компании. http://www.sasolnorthamerica.com/Images/Interior/news/3-12695%20sasol_facts.pdf (на 11.01.2017).
61. Xu J, Yang Y, Li YW. Recent development in converting coal to clean fuels in China. Fuel. 2015; 152:122–130.
62. Pearl GTL. Weblink. http://www.shell.com/about-us/major-projects/pearl-gtl.html (на 11.01.2017).
63. Escravos gas-to-liquid project. Weblink. http://www.hydrocarbons-technology.com/projects/escravos/ (на 11.01.2017).
64. Bartholomew CH. Recent technological developments in Fischer-Tropsch catalysis. Catal. Lett. 1990; 7:303–316.
65. Steynberg A, Dry M (Ed.). Fischer-Tropsch technology. Chapter 3: Chemical concepts used for engineering purposes. Studies Surf. Sci. Catal. 2004; 152:196–257.
66. Steynberg A, Dry M (Ed.). Fischer-Tropsch technology. Chapter 2: Fischer-Tropsch reactors. Studies Surf. Sci. Catal. 2004; 152:64–195.
67. Kopyscinski J, Schildhauer TJ, Biollaz SM. Production of synthetic natural gas (SNG) from coal and dry biomass—A technology review from 1950 to 2009. Fuel. 2010; 8:1763–1783.
68. Li C. Current development situation of Coal to SNG in China. Datang International Chemical Technology Research Institute. 2014. https://www.iea.org/media/workshops/2014/ieamost/3_4_Mr_Li_Chunqi.pdf (на 11.01.2017).
69. Kim JK. Construction Update on SNG Plant in Korea. Gasification Technologies Conference. Washington; 2014 Oct 26–29. http://www.gasification-syngas.org/uploads/eventLibrary/2014_6.2_POSCO_Jin-Kyeong_Kim.pdf (на 11.01.2017).
70. Rönsch S, Schneider J, Matthischke S, Schlüter M, Goetz M, Lefebvre J, Prabhakaran P, Bajohr S. Review on methanation—From fundamentals to current projects. Fuel. 2016; 166:276–296.
71. Rönsch S, Ortwein A. Methanisierung von Synthesegasen - Grundlagen und Verfahrensentwicklungen. Chemie Ingenieur Technik. 2011; 8:1200–1208.
72. Heffer P, Prud'homme H. Fertilizer Outlook 2014-2018. http://www.fertilizer.org/imis20/images/Library_Downloads/2014_ifa_sydney_summary.pdf?WebsiteKey=411e9724-4bda-422f-abfc-8152ed74f306&=404%3bhttp%3a%2f%2fwww.fertilizer.org%3a80%2fen%2fimages%2fLibrary_Downloads%2f2014_ifa_sydney_summary.pdf (на 11.01.2017).
73. China ammonia annual report 2015-2016. http://www.fertmarket.com/newsabout.aspx?id=38 (на 11.01.2017).
74. Appl M. Ammonia. Ullmann's encyclopedia of technical chemistry, 7th ed. Wiley-VCH Verlag GmbH & C. KGaA, Weinheim. 2013.
75. AMIS market monitor. December 2016, Nr. 44, p. 14 http://www.amis-outlook.org/fileadmin/user_upload/amis/docs/Market_monitor/AMIS_Market_Monitor_current.pdf (на 11.01.2017).

76. Moreo P. Casale coal-based ammonia and methanol synloops. Nitrogen & Syngas Conference. Düsseldorf; 2011 Feb 21–24. http://www.casale.ch/images/casalegroup/events/paper_archive/casale_group/2011/nitrogen-syngas_duesseldorf_germany-casale_coal-based_ammonia_and_methanol_synloops.pdf (на 11.01.2017).
77. Finds – a Stokes Engineering Publication. 2013; 28:12–13. http://www.finds.net/The%20U.S.%20Ammonia%20Industry.pdf (на 11.01.2017).
78. Chemical Market Associates Inc. (CMAI), World light olefins analysis. 2009. http://www.ihs.com/products/chemical/index.aspx?pu=1&rd=cmai (на 11.01.2017).
79. Pardemann R. Stoff -Kraft-Kopplung in kohlebasierten Polygenerationkonzepten. Ph.D. thesis; Фрайбергская горная академия, Фрайберг (Германия), 2013.
80. Институт технологий энергопроизводства и химической технологии, Фрайбергская горная академия, Фрайберг (Германия), собственная экономическая оценка, 2016.
81. Tremel A, Wasserscheid P, Baldauf M, Hammer T. Techno-economic analysis for the synthesis of liquid and gaseous fuels based on hydrogen production via electrolysis. International Journal of Hydrogen Energy. 2015; 40:11457–11464.
82. Summers W, Goellner JF, Kuehn NJ, Shah V, White CW, Woods MC. Baseline Analysis of Crude Methanol Production from Coal and Natural Gas. NETL Report, DOE/NETL-341/101514, 15.10.2014. https://www.netl.doe.gov/energy-analyses/temp/BaselineAnalysisofCrudeMethanolProductionfromCoalandNaturalGas_101514.pdf (на 11.01.2017).

Literature

1. Information about the characteristics of Russian coals are provided by the St. Petersburg Mining University.
2. BP Statistical Review of World Energy. 2015. http://www.bp.com/content/dam/bp/pdf/energy-economics/statistical-review-2015/bp-statistical-review-of-world-energy-2015-full-report.pdf (Status 11.01.2017).
3. Добыча топливно-энергетических полезных ископаемых. ГосКомСтат. http://www.gks.ru/free_doc/new_site/business/prom/natura/natura2g.htm (Status 11.01.2017).
4. Mochalnikov SV. Current status and development prospects of coal industry in Russia. International Symposium "Clean Coal Day", Tokyo; 2015 Sept 08–10. http://www.jcoal.or.jp/coaldb/shiryo/material/upload/1-12speech%203%20Russia%20Mr.%20Mochalnikov.pdf (Status 11.01.2017).
5. Higman C, Burgt M. Gasification. Amsterdam, Gulf Professional; 2003.
6. Gräbner M. Industrial coal gasification technologies covering baseline and high-ash coal. Weinheim, Wiley-VCH; 2014.
7. Higman C. State of the gasification industry: worldwide gasification database 2015 update. Colorado Springs; 2015.
8. Timmermann H. Mk Plus—the next Generation Lurgi FBDB Gasification. Leipzig, Germany; 2012.
9. Olschar M, Schulze O. The BGL commercial plants and pilot testing. Leipzig, Germany; 2012.
10. Adlhoch W, Keller J, Herbert PK. The development of the HTW coal gasification process; 1990.
11. Xu J. ECUST coal gasification Update: new projects and operations experience. Washington, D.C.; 2014.
12. Zhuang Q, Bowers K. Unlocking potential value from low rank coal with TRIG gasification technology. Beijing; 2013.
13. Gräbner M, Meyer B. Performance and exergy analysis of the current developments in coal gasification technology. Fuel. 2014; 116:910–20.
14. Laugwitz A, Meyer B. New frontiers and challenges in gasification technologies. In: Nikrityuk PA, Meyer B, editors. Gasification processes: modelling and simulation. Hoboken, Wiley; 2014.
15. Schreiner M. Research guidance studies to assess gasoline from coal by methanol-to-gasoline and Sasol-type Fischer-Tropsch technologies: final report. Mobil R&D Corp. U.S. DoE Report FE-2447-13. Department of Energy; August 1978, 460 p.
16. Methanex Investor Presentation, December 2016. https://www.methanex.com/sites/default/files/MEOH%20Presentation%20Dec%202016.pdf (Status 11.01.2017).
17. Ott J, Gronemann V, Pontzen F, Fiedler E, Grossmann G, Kersebohm B, Weiss G, Witte C. Methanol. Ullmann's encyclopedia of technical chemistry, 7th ed. Wiley-VCH Verlag GmbH & C. KGaA, Weinheim; 2013.

V. Litvinenko and B. Meyer, *Syngas Production: Status and Potential for Implementation in Russian Industry*,
https://doi.org/10.1007/978-3-319-70963-5

18. Higman C, Tam S. Advances in coal gasification, hydrogenation, and gas treating for the production of chemicals and fuels. Chem. Rev. 2014; 114:1673–708.
19. Mastrup M. XtL—the Topsøe approach. 4th International Freiberg Conference on IGCC & XtL Technologies, Dresden; 2010 May 3–6. http://tu-freiberg.de/sites/default/files/media/professur-fuer-energieverfahrenstechnik-und-thermische-rueckstandsbehandlung-16460/publikationen/2010-01-4.pdf (Status 11.01.2017).
20. Minchener AJ. Coal-to-oil, gas and chemicals in China. IEA Clean Coal Center. Report CCC/181. 2011. https://www.usea.org/sites/default/files/022011_Coal-to-oil,%20gas%20and%20chemicals%20in%20China_ccc181.pdf (Status 11.01.2017).
21. Methanol and derivatives. Firmenschrift. https://www.engineering-airliquide.com/sites/activity_eandc/files/2016/07/13/methanol_and_derivatives_brochure-june_2016.pdf (Status 11.01.2017).
22. Successful startup of Hainan Island methanol plant. Weblink. http://www.hydrocarbonprocessing.com/news/2011/02/successful-startup-of-hainan-island-methanol-plant-china (Status 11.01.2017).
23. Tweddle G. Coal & natural gas chemicals. Tecnon OrbiChem Marketing Seminar. Asia Petrochemical Industry Conference, Pattaya; 2014 May 15. http://www.orbichem.com/userfiles/APIC%202014/APIC2014_02_Gillian_Tweddle.pdf (Status 11.01.2017).
24. Methanex monthly average regional posted contract price history. https://www.methanex.com/sites/default/files/methanol-price/MxAvgPrice_Dec%2028%2C%202016.pdf (Status 11.01.2017).
25. Johnson D. Global methanol market review. http://www.ptq.pemex.com/productosyservicios/eventosdescargas/Documents/Foro%20PEMEX%20Petroqu%C3%ADmica/2012/PEMEX_DJohnson.pdf (Status 11.01.2017).
26. Methanol synthesis, the synthesis loop. SUPERMETHANOL Project in the framework of the 7th European Commission framework programme. http://www.supermethanol.eu/index.php?id=22&rid=13&r=methanol_synthesis (Status 11.01.2017).
27. Chen L, Jiang Q, Song Z, Posarac D. Optimization of methanol yield from Lurgi reactor. Chem. Engineering Techn. 2011; 34:817–22.
28. Haid J, Koss U. Lurgi's mega methanol technology opens the door for a new era in down-stream applications. Studies Surf. Sci. Catal. 2001; 136:399–404.
29. Wurzel T. Lurgi MegaMethanol technology—delivering the building blocks for future fuel and monomer demand. DGMK Conference "Synthesis Gas Chemistry", Dresden; 2006 Oct 04–06. http://www.dgmk.de/petrochemistry/abstracts_content14/Wurzel.pdf (Status 11.01.2017).
30. Heydorn EC, Diamond BW, Lilly RD. Commercial scale demonstration of the liquid phase methanol (LPMeOH) process. Final report DoE/NETL: DE-FC22-92PC90543. June 2003. https://www.netl.doe.gov/File%20Library/Research/Coal/major%20demonstrations/cctdp/Round3/LPMEOH/LPMEOHFinal.pdf (Status 11.01.2017).
31. Air Products and Chemicals Inc. Economic Analysis: LPMeOH process as an Add-on Integrated Gasification Combined Cycle for coproduction. Topical report DoE/NETL: DE-FC22-92PC90543. October 1998. https://www.netl.doe.gov/File%20Library/Research/Coal/major%20demonstrations/cctdp/Round3/LPMEOH/Econfin.pdf (Status 11.01.2017).
32. Chang CL, Chu CTW, Socha RF. Methanol conversion to olefins over ZSM-5. I. Effect of temperature and zeolite SiO_2/Al_2O_3. J. Catal. 1984; 86:289–96.
33. Chang CL, Silvestri AJ. MTG. Origin, evolution, operation. Chem. Tech. 1987; 17:624–31.
34. Keil FJ. Methanol-to-hydrocarbon: process technology. Micropor. Mesopor. Mater. 1999; 29:49–66.
35. Topp-Jorgensen J, Topsøe integrated gasoline synthesis—the TIGAS process. Studies Surf. Sci. Catal. 1988; 36:293–305.
36. Kuschel M, Engelmann J, Meyer B. STF technology—production of high octane gasoline from synthesis gas on a new route. International Pittsburgh Coal Conference, Beijing; 2013 Sept 16–18.

37. Baitalow F, Stahlschmidt R, Seifert P, Pardemann R, Meyer B, Engelmann J. Production of gasoline from coal based on the STF process. 18th International Coal Preparation Congress, St. Petersburg; 2016 June 28–Jul 1.
38. Bertau M. Offermanns H, Plass L, Schmidt F, Wernicke HJ (ed.). Methanol: the basic chemical and energy feedstock of the future. Springer, Heidelberg; 2014.
39. China's Yulin Energy to start up new MTO unit in June 2014. Weblink. www.icis.com/resources/news/2013/10/10/9713958/china-s-yulin-energy-to-start-up-new-mto-unit-in-june-2014/ (Status 11.01.2017).
40. KBR deploys its methanol-to-olefin recovery technology in China. Weblink. www.2b1stconsulting.com/kbr-deploys-its-methanol-to-olefin-recovery-technology-in-china/ (Status 11.01.2017).
41. Chemiewirtschaft in Zahlen. VCI. August 2016. https://www.vci.de/vci/downloads-vci/publikation/chemiewirtschaft-in-zahlen-print.pdf (Status 11.01.2017).
42. Chang CL. Hydrocarbons from methanol. Catal. Rev. Sci. Eng. 1983:1–118.
43. Grimmer HR, Thiagarajan N, Nitschke E. Conversion of methanol to liquid fuels by the fluid bed Mobil process (a commercial concept). Studies Surf. Sci. Catal. 1988; 36:273–91.
44. Gould RM, Tabak SA. Multi-stage system for converting oxygenates to liquid hydrocarbons with aliphatic recycle. U.S. Patent 4,689,205 A. 1987.
45. Tabak SA, Avidan AA, Krambeck FJ. Production of synthetic gasoline and diesel fuel from non-petroleum resources. ACS National Meeting. New York; 1986 April 13–18. https://web.anl.gov/PCS/acsfuel/preprint%20archive/Files/31_2_NEW%20YORK_04-86_0293.pdf (Status 11.01.2017).
46. Liebner W, Wagner M. MtSynfuels, efficient and economical alternative to Fischer-Tropsch fuels. Erdöl, Erdgas, Kohle. 2004; 120:323–26.
47. Rothaemel M, Holtmann H.-D. Methanol to propylene MTP—Lurgi's way. Oil Gas European Magazine. 2002; 27–30.
48. Koempel H, Liebner W. Lurgi's Methanol to propylene (MTP): report on a successful commercialisation. Studies Surf. Sci. Catal. 2007; 167:261–67.
49. Wurzel T, Koempel H, Rothaemel M, Liebner W. MTP—the on-purpose propylene technology for a changing feedstock environment. 19th World Petroleum Congress. Madrid; 2008 June 29–Jul 03.
50. Vora BV, Marker TL, Barger PT, Nilsen HR, Kvisle S, Fuglerud T. Economic route for natural gas conversion to ethylene and propylene. Studies Surf. Sci. Catal. 1988; 36:87–98.
51. Chen JQ, Bozzano A, Glover B, Fuglerud T, Kvisle S. Recent advancements in ethylene and propylene production using the UOP/Hyro MTO process. Catal. Today. 2005; 106:103–07.
52. Quann RJ, Green LA, Tabak SA, Krambeck FJ. Chemistry of olefin oligomerization over ZSM-5 catalyst. Ind. Eng. Chem. Res. 1988; 27:565–70.
53. Tabak SA, Krambeck FJ. Shaping process making fuels. Hydrocarbon Processing. 1985; 64:72–74.
54. Tabak SA. Production of synthetic diesel fuel from light olefins. AIChE National Meeting. Philadelphia; 1984 Aug 26–31.
55. Koempel H, Liebner W, Wagner M. Lurgi's gas to chemicals (GTC): advanced technologies for natural gas monetisation. Gastech. Bilbao; 2005 Mar 14–17. http://www.ivt.ntnu.no/ept/fag/tep4215/innhold/LNG%20Conferences/2005/SDS_TIF/050140.pdf (Status 11.01.2017).
56. Koehler E, Schmidt F, Wernicke HJ, de Pontes M, Roberts HL. Converting olefins to diesel —the COD process. Hydrocarbon Techn. Int. Summer. 1995:37–40.
57. Knottenbelt CD, Dunlop C, Zono K. Process for catalytic conversion of Fischer-Tropsch derived olefins into distillates. U.S. Patent 8,318,003 B2. 2012.
58. Knottenbelt CD, Ntshabele JPP. PetroSA's conversion of olefins to distillate (COD) research program. 18th World Petroleum Congress. Johannesburg; 2005 Sept 25–29.

59. Steynberg A, Dry M (Ed.). Fischer-Tropsch technology. Chapter 5: Commercial Fischer-Tropsch process application. Studies Surf. Sci. Catal. 2004; 152:406–81.
60. SASOL Facts 2012/13. Company brochure. http://www.sasolnorthamerica.com/Images/Interior/news/3-12695%20sasol_facts.pdf (Status 11.01.2017).
61. Xu J, Yang Y, Li YW. Recent development in converting coal to clean fuels in China. Fuel. 2015; 152:122–30.
62. Pearl GTL. Weblink. http://www.shell.com/about-us/major-projects/pearl-gtl.html (Status 11.01.2017).
63. Escravos gas-to-liquid project. Weblink. http://www.hydrocarbons-technology.com/projects/escravos/ (Status 11.01.2017).
64. Bartholomew CH. Recent technological developments in Fischer-Tropsch catalysis. Catal. Lett. 1990; 7:303–16.
65. Steynberg A, Dry M (Ed.). Fischer-Tropsch technology. Chapter 3: Chemical concepts used for engineering purposes. Studies Surf. Sci. Catal. 2004; 152: 196–257.
66. Steynberg A, Dry M (Ed.). Fischer-Tropsch technology. Chapter 2: Fischer-Tropsch reactors. Studies Surf. Sci. Catal. 2004; 152:64–195.
67. Kopyscinski J, Schildhauer TJ, Biollaz SM. Production of synthetic natural gas (SNG) from coal and dry biomass—a technology review from 1950 to 2009. Fuel. 2010; 8:1763–83.
68. Li C. Current development situation of coal to SNG in China. Datang International Chemical Technology Research Institute. 2014. https://www.iea.org/media/workshops/2014/ieamost/3_4_Mr_Li_Chunqi.pdf (Status 11.01.2017).
69. Kim JK. Construction Update on SNG Plant in Korea. Gasification Technologies Conference. Washington; 2014 Oct 26–29. http://www.gasification-syngas.org/uploads/eventLibrary/2014_6.2_POSCO_Jin-Kyeong_Kim.pdf (Status 11.01.2017).
70. Rönsch S, Schneider J, Matthischke S, Schlüter M, Goetz M, Lefebvre J, Prabhakaran P, Bajohr S. Review on methanation—from fundamentals to current projects. Fuel. 2016; 166:276–96.
71. Rönsch S, Ortwein A. Methanisierung von Synthesegasen - Grundlagen und Verfahrensentwicklungen. Chemie Ingenieur Technik. 2011; 8:1200–08.
72. Heffer P, Prud'homme H. Fertilizer Outlook 2014–2018. http://www.fertilizer.org/imis20/images/Library_Downloads/2014_ifa_sydney_summary.pdf?WebsiteKey=411e9724-4bda-422f-abfc-8152ed74f306&=404%3bhttp%3a%2f%2fwww.fertilizer.org%3a80%2fen%2fimages%2fLibrary_Downloads%2f2014_ifa_sydney_summary.pdf (Status 11.01.2017).
73. China ammonia annual report 2015–2016. http://www.fertmarket.com/newsabout.aspx?id=38 (Status 11.01.2017).
74. Appl M. Ammonia. Ullmann's encyclopedia of technical chemistry, 7th ed. Wiley-VCH Verlag GmbH & C. KGaA, Weinheim; 2013.
75. AMIS market monitor. December 2016, Nr. 44, p. 14. http://www.amis-outlook.org/fileadmin/user_upload/amis/docs/Market_monitor/AMIS_Market_Monitor_current.pdf (Status 11.01.2017).
76. Moreo P. Casale coal-based ammonia and methanol synloops. Nitrogen & Syngas Conference. Düsseldorf; 2011 Feb 21–24. http://www.casale.ch/images/casalegroup/events/paper_archive/casale_group/2011/nitrogen-syngas_duesseldorf_germany-casale_coal-based_ammonia_and_methanol_synloops.pdf (Status 11.01.2017).
77. Finds – a Stokes Engineering Publication. 2013; 28:12–13. http://www.finds.net/The%20U.S.%20Ammonia%20Industry.pdf (Status 11.01.2017).
78. Chemical Market Associates Inc. (CMAI), World light olefins analysis. 2009. http://www.ihs.com/products/chemical/index.aspx?pu=1&rd=cmai (Status 11.01.2017).
79. Pardemann R. Stoff -Kraft-Kopplung in kohlebasierten Polygenerationkonzepten. Ph.D. thesis; TU Bergakademie. Freiberg; 2013.
80. IEC. Internal cost calculations. 2016.

81. Tremel A, Wasserscheid P, Baldauf M, Hammer T. Techno-economic analysis for the synthesis of liquid and gaseous fuels based on hydrogen production via electrolysis. International Journal of Hydrogen Energy. 2015; 40:11457–64.
82. Summers W, Goellner JF, Kuehn NJ, Shah V, White CW, Woods MC. Baseline analysis of crude methanol production from coal and natural gas. NETL Report, DOE/NETL-341/101514, 15.10.2014. https://www.netl.doe.gov/energy-analyses/temp/BaselineAnalysisofCrudeMethanolProductionfromCoalandNaturalGas_101514.pdf (Status 11.01.2017).

MIX
Papier aus verantwortungsvollen Quellen
Paper from responsible sources
FSC® C105338

If you have any concerns about our products,
you can contact us on
ProductSafety@springernature.com

In case Publisher is established outside the EU,
the EU authorized representative is:
Springer Nature Customer Service Center GmbH
Europaplatz 3, 69115 Heidelberg, Germany

Printed by Libri Plureos GmbH
in Hamburg, Germany